SCIENCE STARTERS:
ELEMENTARY GENERAL SCIENCE & ASTRONOMY

Investigate the Possibilities

Elementary General Science

Water & Weather
From the Flood to Forecasts

Tom DeRosa
Carolyn Reeves

Elementary Astronomy

The Universe
From Comets to Constellations

Tom DeRosa
Carolyn Reeves

P.L.P.
Parent Lesson Planner

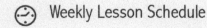 Weekly Lesson Schedule

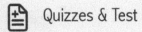 Quizzes & Test

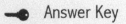 Answer Key

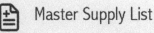 Master Supply List

First printing: April 2014
Second printing: June 2015

ISBN: 978-0-89051-810-6

Printed in the United States of America

Please visit our website for other great titles:
www.masterbooks.net

For information regarding author interviews,
please contact the publicity department at (870) 438-5288

Master Books®
A Division of New Leaf Publishing Group
www.masterbooks.net

Where Creation Inspires Education

Since 1975, Master Books has been providing educational resources based on a biblical worldview to students of all ages. At the heart of these resources is our firm belief in a literal six-day creation, a young earth, the global Flood as revealed in Genesis 1–11, and other vital evidence to help build a critical foundation of scriptural authority for everyone. By equipping students with biblical truths and their key connection to the world of science and history, it is our hope they will be able to defend their faith in a skeptical, fallen world.

If the foundations are destroyed, what can the righteous do?
Psalm 11:3

As the largest publisher of creation science materials in the world, Master Books is honored to partner with our authors and educators, including:

Ken Ham of Answers in Genesis

Dr. John Morris and Dr. Jason Lisle of the Institute for Creation Research

Dr. Donald DeYoung and Michael Oard of the Creation Research Society

Dr. James Stobaugh, John Hudson Tiner, Rick and Marilyn Boyer, Dr. Tom DeRosa, Todd Friel, Israel Wayne, Michael Farris, and so many more!

Whether a pre-school learner or a scholar seeking an advanced degree, we offer a wonderful selection of award-winning resources for all ages and educational levels.

But sanctify the Lord God in your hearts, and always be ready
to give a defense to everyone who asks you a reason for the hope
that is in you, with meekness and fear.
1 Peter 3:15

Permission to Copy

Contents

Lessons for a 36-week course!

Overview: This *Science Starters PLP* contains materials for use with *Investigate the Possibilities: Water & Weather – From Floods to Forecasts* and *Investigate the Possibilities: The Universe – From Comets to Constellations.* Materials are organized by each book in the following sections:

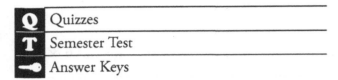

Q	Quizzes
T	Semester Test
🔑	Answer Keys

Multi-level Quiz & Test Options: The Science Starters curriculum allows multi-age students between grades 3 and 6 to be taught at the same time. Worksheets and quizzes are perforated and three-hole punched — materials are easy to tear out, hand out, grade, and store. For your convenience, we have included two different levels of quizzes and semester tests so that you can choose the ones most appropriate for your student's age and educational abilities. Suggested levels include:

Level 1 – Grades 3 to 4
Level 2 – Grades 5 and 6

Workflow: Students will read two pages in their book and then complete one page of the Student Journal. Extra Projects are also assigned. Quizzes are given at regular intervals.

Lesson Scheduling: Space is given for assignment dates. There is flexibility in scheduling. While each quarter has 45 days of assignments, they do not have to be given M-F. Some students may prefer to do more assignments each day, allowing for breaks on other days. Each week listed has five days, but due to vacations the school work week may not be M-F. Please adapt the days to your school schedule. As the student completes each assignment, he/she should put an "X" in the box.

🕐	Approximately 30 to 45 minutes per lesson, two to three days a week.	Course includes books from creationist authors with solid, biblical worldviews:
🔑	Includes answer keys for quizzes and semester test.	**Tom DeRosa** is an experienced science educator, a committed creationist, and founder/director of a growing national creation organization whose chief focus is education. His experience in the public school, Christian school, and homeschool markets for over 35 years has given him special insights into what really works in engaging young minds. He holds a master's degree in education, with the emphasis of science curriculum.
🔁	Multi-level quizzes and tests are included to help reinforce learning and provide assessment opportunities.	**Carolyn Reeves** is especially skilled at creating ways to help students develop a greater understanding of not just scientific concepts, but also how these are applied within the world around us. Carolyn retired after a 30-year career as a science teacher, finished a doctoral degree in science education, and now serves as a writer and an educational consultant.
📄	Designed for grades 3 to 6 in a one-year course. Master supply list included.	

Science Starters: Physical and Earth Science

Course Description

This is the suggested course sequence that allows one core area of science to be studied per semester. You can change the sequence of the semesters per the needs or interests of your student; materials for each semester are independent of one another to allow flexibility.

Semester 1: Water and Weather

Learn about zones in the ocean, from warm, clear water to the deepest, coldest places. Find out the facts about the weather cycle, earth's purification system, weather instruments, and more! Discover the important connection between water and fossils—how this helped to form, alter, and reveal them. Study not only the weather and varying climates around the world, but also explore the results of weather events in the past. The great Flood was a cataclysmic event that left behind fossils, and these impressions reveal much about what happened during and after this historical event. Form a connection between this biblical history and the world experienced outside your door as natural processes like rain and wind are shown to be forces at work in the environment!

Semester 2: The Universe

How big is the solar system? How big is the universe? Can we make a model to help us understand God's wonderful creation? These and other questions are answered through a fun and investigative process created just for upper elementary students! Through simple experiments and fact-finding problems, this astronomy curriculum brings to light God's design of this massive and intricate universe. Students will read about the historical discoveries of great scientists of the past like Kepler, Galileo, and Newton, and how their words impact us today. They will explore astronomy from the first observations of space, the creation of the telescope, the history of flight, and more. Students are encouraged in their faith as they become engaged in the scientific process through activities using inexpensive, everyday household items that bring science to life.

Calculating a Final Grade

Calculate the Average of the student's Activities & Observations grades.

               Divide the average by 3   _____

Calculate the Average of the student's Questions & Quizzes grades.

               Divide the average by 3   _____

Calculate the Average of the student's Projects, Contest & Dig Deeper grades.

               Divide the average by 3   _____

Add up the numbers for the Final Grade: _____

First Semester Suggested Daily Schedule

Date	Day	Assignment	Due Date	✓	Grade
		First Semester-First Quarter — *Water & Weather*			
Week 1	Day 1				
	Day 2	Investigation #1: In the Beginning...God Created Dinosaurs! Read Pages 4-7 • *Water & Weather* (WW) Complete Page S4 • Student Journal (SJ)			
	Day 3				
	Day 4	Investigation #1: In the Beginning...God Created Dinosaurs! Read Pages 8-9 • (WW) • Complete Page S5 • (SJ)			
	Day 5				
Week 2	Day 6				
	Day 7	Investigation #2: Making a Big Impression! Read Pages 10-11 • (WW) • Complete Page S6 • (SJ)			
	Day 8				
	Day 9	Investigation #2: Making a Big Impression! Read Pages 12-13 • (WW) • Complete Page S7 • (SJ)			
	Day 10				
Week 3	Day 11				
	Day 12	Investigation #3: No Bones about It! Read Pages 14-15 • (WW) • Complete Page S8 • (SJ)			
	Day 13				
	Day 14	Investigation #3: No Bones about It! Read Pages 16-17 • (WW) • Complete Page S9 • (SJ)			
	Day 15				
Week 4	Day 16	Investigation #4: Digging in and Reconstructing Fossils Read Pages 18-19 • (WW) • Complete Page S10 • (SJ)			
	Day 17				
	Day 18	Investigation #4: Digging in and Reconstructing Fossils Read Pages 20-21 • (WW) • Complete Page S11 • (SJ)			
	Day 19				
	Day 20	**Water & Weather Investigations 1-4 Quiz 1** **Level 1** Page 17 • **Level 2** Page 33 • Lesson Plan (LP)			
Week 5	Day 21	Investigation #5: Can Rocks Tell Time? Read Pages 22-23 • (WW) • Complete Page S12 • (SJ)			
	Day 22				
	Day 23	Investigation #5: Can Rocks Tell Time? Read Pages 24-25 • (WW) • Complete Page S13 • (SJ)			
	Day 24				
	Day 25	Investigation #6: Leave No Stone Unturned Read Pages 26-27 • (WW) • Complete Page S14 • (SJ)			

Date	Day	Assignment	Due Date	✓	Grade
Week 6	Day 26	Investigation #6: Leave No Stone Unturned Read Pages 28-29 • (WW) • Complete Page S15 • (SJ)			
	Day 27				
	Day 28	**Water & Weather Investigations 5-6 Quiz 2** **Level 1** Page 19 • **Level 2** Page 35 • (LP)			
	Day 29				
	Day 30	Investigation #7: Just How Salty Is the Ocean? Read Pages 30-31 • (WW) • Complete Page S16 • (SJ)			
Week 7	Day 31	Investigation #7: Just How Salty Is the Ocean? Read Pages 32-33 • (WW) • Complete Page S17 • (SJ)			
	Day 32				
	Day 33	Investigation #8: Ocean Zones — From Light to Dark Places Read Pages 34-35 • (WW) • Complete Page S18 • (SJ)			
	Day 34				
	Day 35	Investigation #8: Ocean Zones — From Light to Dark Places Read Pages 36-37 • (WW) • Complete Page S19 • (SJ)			
Week 8	Day 36	Investigation #9: From Shelf to Shelf and In-Between Read Pages 38-39 • (WW) • Complete Page S20 • (SJ)			
	Day 37				
	Day 38	Investigation #9: From Shelf to Shelf and In-Between Read Pages 40-41 • (WW) • Complete Page S21 • (SJ)			
	Day 39				
	Day 40	**Water & Weather Investigations 7-9 Quiz 3** **Level 1** Page 21 • **Level 2** Page 37 • (LP)			
Week 9	Day 41				
	Day 42	Investigation #10: Currents in the Ocean Read Pages 42-43 • (WW) • Complete Page S22 • (SJ)			
	Day 43				
	Day 44	Investigation #10: Currents in the Ocean Read Pages 44-45 • (WW) • Complete Page S23 • (SJ)			
	Day 45				
First Semester-Second Quarter — *Water & Weather*					
Week 1	Day 46	Investigation #11: Where the Rivers Run Into the Sea Read Pages 46-47 • (WW) • Complete Page S24 • (SJ)			
	Day 47				
	Day 48	Investigation #11: Where the Rivers Run Into the Sea Read Pages 48-49 • (WW) • Complete Page S25 • (SJ)			
	Day 49				
	Day 50	Investigation #12: Evaporation, Condensation & the Water Cycle Read Pages 50-51 • (WW) • Complete Page S26 • (SJ)			

Date	Day	Assignment	Due Date	✓	Grade
	Day 51	Investigation #12: Evaporation, Condensation & the Water Cycle Read Pages 52-53 • (WW) • Complete Page S27 • (SJ)			
	Day 52				
Week 2	Day 53	Investigation #13: God's System of Purifying Water Read Pages 54-55 • (WW) • Complete Page S28 • (SJ)			
	Day 54				
	Day 55	Investigation #13: God's System of Purifying Water Read Pages 56-57 • (WW) • Complete Page S29 • (SJ)			
	Day 56	**Water & Weather Investigations 10-13 Quiz 4** **Level 1** Page 23 • **Level 2** Page 39 • (LP)			
	Day 57				
Week 3	Day 58	Investigation #14: Weather or Not! Read Pages 58-59 • (WW) • Complete Page S30 • (SJ)			
	Day 59				
	Day 60	Investigation #14: Weather or Not! Read Pages 60-61 • (WW) • Complete Page S31 • (SJ)			
	Day 61				
	Day 62	Investigation #15: Day and Night, Summer and Winter Read Pages 62-63 • (WW) • Complete Page S32 • (SJ)			
Week 4	Day 63				
	Day 64	Investigation #15: Day and Night, Summer and Winter Read Pages 64-65 • (WW) • Complete Page S33 • (SJ)			
	Day 65				
	Day 66	Investigation #16: The Temperature's Rising Read Pages 66-67 • (WW) • Complete Page S34 • (SJ)			
	Day 67				
Week 5	Day 68	Investigation #16: The Temperature's Rising Read Pages 68-69 • (WW) • Complete Page S35 • (SJ)			
	Day 69				
	Day 70	**Water & Weather Investigations 14-16 Quiz 5** **Level 1** Page 25 • **Level 2** Page 41 • (LP)			
	Day 71				
	Day 72	Investigation #17: Weather Instruments Read Pages 70-72 • (WW) • Complete Page S36 • (SJ)			
Week 6	Day 73				
	Day 74	Investigation #17: Weather Instruments Read Pages 73-75 • (WW) • Complete Page S37 • (SJ)			
	Day 75				

Date	Day	Assignment	Due Date	✓	Grade
Week 7	Day 76	Investigation #18: Forecasting the Weather Read Pages 76-77 • (WW) • Complete Page S38 • (SJ)			
	Day 77				
	Day 78	Investigation #18: Forecasting the Weather Read Pages 78-79 • (WW) • Complete Page S39 • (SJ)			
	Day 79				
	Day 80	Investigation #19: From Gentle Breezes to Dangerous Winds Read Pages 80-81 • (WW) • Complete Page S40 • (SJ)			
Week 8	Day 81	Investigation #19: From Gentle Breezes to Dangerous Winds Read Pages 82-83 • (WW) • Complete Page S41 • (SJ)			
	Day 82				
	Day 83	Investigation #20: Climate Change Read Pages 84-85 • (WW) • Complete Page S42 • (SJ)			
	Day 84				
	Day 85	Investigation #20: Climate Change Read Pages 86-87 • (WW) • Complete Page S43 • (SJ)			
Week 9	Day 86				
	Day 87	**Water & Weather Investigations 17-20 Quiz 6** **Level 1** Page 27 • **Level 2** Page 43 • (LP)			
	Day 88				
	Day 89	**Water & Weather Investigations 1-20 Test** **Level 1** Page 29 • **Level 2** Page 45 • (LP)			
	Day 90				
		Mid-Term Grade			

Second Semester Suggested Daily Schedule

Date	Day	Assignment	Due Date	✓	Grade
		Second Semester-Third Quarter — *The Universe*			
Week 1	Day 91				
	Day 92	Investigation #1: What Is the Universe? Read Pages 4-7 • *The Universe* (TU) Complete Page S4 • Student Journal (SJ)			
	Day 93				
	Day 94	Investigation #1: What Is the Universe? Read Pages 8-9 • (TU) • Complete Page S5 • (SJ)			
	Day 95				
Week 2	Day 96				
	Day 97	Investigation #2: Spreading Out the Heavens Read Pages 10-11 • (TU) • Complete Page S6 • (SJ)			
	Day 98				
	Day 99	Investigation #2: Spreading Out the Heavens Read Pages 12-13 • (TU) • Complete Page S7 • (SJ)			
	Day 100				
Week 3	Day 101	Investigation #3: The Strange Behavior of Space and Light Read Pages 14-15 • (TU) • Complete Page S8 • (SJ)			
	Day 102				
	Day 103	Investigation #3: The Strange Behavior of Space and Light Read Pages 16-17 • (TU) • Complete Page S9 • (SJ)			
	Day 104				
	Day 105	**The Universe Investigations 1-3 Quiz 1** **Level 1** Page 51 • **Level 2** Page 65 • **Lesson Plan** • (LP)			
Week 4	Day 106	Investigation #4: Kepler's Clockwise Universe Read Pages 18-19 • (TU) • Complete Page S10 • (SJ)			
	Day 107				
	Day 108	Investigation #4: Kepler's Clockwise Universe Read Pages 20-21 • (TU) • Complete Page S11 • (SJ)			
	Day 109				
	Day 110	Investigation #5: Invisible Forces in the Universe Read Pages 22-23 • (TU) • Complete Page S12 • (SJ)			
Week 5	Day 111	Investigation #5: Invisible Forces in the Universe Read Pages 24-25 • (TU) • Complete Page S13 • (SJ)			
	Day 112				
	Day 113	Investigation #6: Galileo and Inertia Read Pages 26-27 • (TU) • Complete Page S14 • (SJ)			
	Day 114				
	Day 115	Investigation #6: Galileo and Inertia Read Pages 28-29 • (TU) • Complete Page S15 • (SJ)			

Date	Day	Assignment	Due Date	✓	Grade
Week 6	Day 116	Investigation #7: Making Telescopes Read Pages 30-31 • (TU) • Complete Page S16 • (SJ)			
	Day 117				
	Day 118	Investigation #7: Making Telescopes Read Pages 32-33 • (TU) • Complete Page S17 • (SJ)			
	Day 119				
	Day 120	**The Universe Investigations 4-7 Quiz 2** **Level 1** Page 53 • **Level 2** Page 67 • (LP)			
Week 7	Day 121	Investigation #8: History of Flight Read Pages 34-35 • (TU) • Complete Page S18 • (SJ)			
	Day 122				
	Day 123	Investigation #8: History of Flight Read Pages 36-37 • (TU) • Complete Page S19 • (SJ)			
	Day 124				
	Day 125	Investigation #9: Rockets and Space Exploration Read Pages 38-39 • (TU) • Complete Page S20 • (SJ)			
Week 8	Day 126	Investigation #9: Rockets and Space Exploration Read Pages 40-41 • (TU) • Complete Page S21 • (SJ)			
	Day 127				
	Day 128	Investigation #10: The Earth in Space Read Pages 42-43 • (TU) • Complete Page S22 • (SJ)			
	Day 129				
	Day 130	Investigation #10: The Earth in Space Read Pages 44-45 • (TU) • Complete Page S23 • (SJ)			
Week 9	Day 131	Investigation #11: The Earth's Atmosphere Read Pages 46-47 • (TU) • Complete Page S24 • (SJ)			
	Day 132				
	Day 133	Investigation #11: The Earth's Atmosphere Read Pages 48-49 • (TU) • Complete Page S25 • (SJ)			
	Day 134				
	Day 135	**The Universe 8-11 Quiz 3** **Level 1** Page 55 • **Level 2** Page 69 • (LP)			
Second Semester-Fourth Quarter — *The Universe*					
Week 1	Day 136				
	Day 137	Investigation #12: The Moon Read Pages 50-51 • (TU) • Complete Page S26 • (SJ)			
	Day 138				
	Day 139	Investigation #12: The Moon Read Pages 52-53 • (TU) • Complete Page S27 • (SJ)			
	Day 140				

Date	Day	Assignment	Due Date	✓	Grade
	Day 141	Investigation #13: Rocky Planets: Mercury, Venus, Earth & Mars Read Pages 54-55 • (TU) • Complete Page S28 • (SJ)			
	Day 142				
Week 2	Day 143	Investigation #13: Rocky Planets: Mercury, Venus, Earth & Mars Read Pages 56-57 • (TU) • Complete Page S29 • (SJ)			
	Day 144				
	Day 145	Investigation #14: Planets: Mars & Martians Read Pages 58-59 • (TU) • Complete Page S30 • (SJ)			
	Day 146	Investigation #14: Planets: Mars & Martians Read Pages 60-61 • (TU) • Complete Page S31 • (SJ)			
	Day 147				
Week 3	Day 148	Investigation #15: Jovian Planets: Jupiter, Saturn, Uranus & Neptune Read Pages 62-63 • (TU) • Complete Page S32 • (SJ)			
	Day 149				
	Day 150	Investigation #15: Jovian Planets: Jupiter, Saturn, Uranus & Neptune Read Pages 64-65 • (TU) • Complete Page S33 • (SJ)			
	Day 151	**The Universe Investigations 12-15 Quiz 4** **Level 1** Page 57 • **Level 2** Page 71 • **(LP)**			
	Day 152				
Week 4	Day 153	Investigation #16: The Sun and Its Light Read Pages 66-67 • (TU) • Complete Page S34 • (SJ)			
	Day 154				
	Day 155	Investigation #16: The Sun and Its Light Read Pages 68-69 • (TU) • Complete Page S35 • (SJ)			
	Day 156				
	Day 157	Investigation #17: The Sun and the Earth Relationship Read Pages 70-71 • (TU) • Complete Page S36 • (SJ)			
Week 5	Day 158				
	Day 159	Investigation #17: The Sun and the Earth Relationship Read Pages 72-73 • (TU) • Complete Page S37 • (SJ)			
	Day 160				
	Day 161				
	Day 162	Investigation #18: The Constellations Read Pages 74-75 • (TU) • Complete Page S38 • (SJ)			
Week 6	Day 163				
	Day 164	Investigation #18: The Constellations Read Pages 76-77 • (TU) • Complete Page S39 • (SJ)			
	Day 165				

Date	Day	Assignment	Due Date	✓	Grade
Week 7	Day 166				
	Day 167	Investigation #19: A Great Variety in Space Read Pages 78-79 • (TU) • Complete Page S40 • (SJ)			
	Day 168				
	Day 169	Investigation #19: A Great Variety in Space Read Pages 80-81 • (TU) • Complete Page S41 • (SJ)			
	Day 170				
Week 8	Day 171				
	Day 172	Investigation #20: Chaos or a Creator? Read Pages 82-83 • (TU) • Complete Page S42 • (SJ)			
	Day 173				
	Day 174	Investigation #20: Chaos or a Creator? Read Pages 84-85 • (TU) • Complete Page S43 • (SJ)			
	Day 175				
Week 9	Day 176				
	Day 177	**The Universe 16-20 Quiz 5** **Level 1** Page 59 • **Level 2** Page 73 • (LP)			
	Day 178				
	Day 179				
	Day 180	**The Universe Investigations 1-20 Test** **Level 1** Page 61 • **Level 2** Page 75 • (LP)			
		Final Grade			

Quizzes and Test

for Use with

Water and Weather

Testing:

This series is appropriate for elementary students from 3rd to 6th grades. Because of this, we have included quizzes and tests in two different levels, which you can choose from based on your child's abilities and understanding of the concepts in the course.

Level 1: suggested for younger ages or those who struggle with application of the concepts beyond just definitions and basic concepts

Level 2: suggested for older ages or those who can both grasp the scientific concepts and apply them consistently

Permission to Copy

Choose answers from these terms.
All the terms may not be used and some may be used more than once:

decay	sedimentary	no	index fossils	superposition
geologic column	yes	magma	unconformities	oldest
worldwide flood	graveyards	evolutionary	radioactive	half life
daughter	bottom	erosion	uplift	assumptions

Fill in the Blank: Each question is worth 5 points.

1. In order to reach a more stable state, some radioactive atoms throw out little bits of nuclear particles, causing the original element to change into something else. This process is known as radioactive _____.

2. The length of time it takes for one half of the original radioactive element to change into _____ elements is called its _____ _____.

3. _____ rock is composed of many rocks of different ages that were cemented together.

4. Evolutionary scientists usually attempt to date sedimentary rocks by _____ _____ found in the rocks, by dating igneous rocks around them, or by how they fit into the geologic column.

5. The principle of superposition states that the _____ layers, along with fossils and artifacts in them, are on the _____.

6. Evidence of _____ between layers are called unconformities.

7. Intrusive rock formations may form as liquid _____ from deep below the ground squeezes through cracks in existing sedimentary rocks.

8. Creation scientists believe that the mass extinction of plants and animals came about because of a _____ _____.

9. Fossil _____ occur when plants and animals from one environmental location may have been transported to another location.

10. The geologic column was established and agreed upon according to _____ ideas.

11. Does absolute dating mean the same thing as proven correct? _____

True / False: Each answer is worth 5 points.

12. During the process of radioactive decay, one element changes into another.

13. The amount of the original radioactive element becomes more and more over time.

14. The rate at which radioactive decay occurs is not affected by temperature, pressure, or other chemicals.

15. When radioactive dating methods are used to date the age of certain rocks, they are always correct.

16. The ages of antiques and certain pieces of art are sometimes obtained by using carbon 14 dating.

17. Radiometric dating is only accurate if the assumptions that are made are accurate.

18. There are very few places on earth where all the layers of the geologic column are found in the "official" order.

*Bonus Question: 10 points

19. The surface of the earth's continents consists mostly of sedimentary rock layers. These layers often extend over several states, sometimes over entire countries! How does water play a part in the formation of these layers?

*Bonus questions will be included in the test at the end of the semester.

Choose answers from these terms.
All the terms may not be used and some may be used more than once:

layers	fog	prevailing winds	tilt	satellites
thermometer	cirrus	slanted	revolution	humid
cooling	cumulus	stratus	rain	dry
dew	frost	heating	weight	tropical
cumulonimbus	layers	arctic	dispersed	more
rotation	concentrated	less	equator	hemisphere

Fill in the Blank: Each question is worth 5 points.

1. _____ clouds are thin and wispy.

2. _____ clouds look white and fluffy.

3. Water droplets that float near the ground are called _____.

4. Clouds form when warm, _____ air rises and condenses into tiny droplets of water.

5. Nimbus means _____.

6. Stratus means _____.

7. _____ are the tallest of all clouds.

8. _____ is similar to fog, but the water vapor condenses on the ground instead of in the air.

9. The _____ of the earth is what causes the seasons.

10. The most _____ rays of the sun strike an area in the morning or the evening.

11. When the earth completes a path around the sun it is called a _____.

12. _____ regions of the earth receive more direct rays from the sun than other regions.

13. A basic understanding of the weather can be obtained by understanding the effects of two main principles; the uneven _____ of the earth and the _____ of the air.

14. _____ _____ over the United States tend to blow from the west to the east.

15. _____ in space provide an amazing amount of weather information for large areas of the globe at a time.

16. Direct rays from the sun are more _____ than slanted rays, and therefore produce _____ heat.

True / False: Each answer is worth 5 points.

17. When air molecules are heated, the increased heat energy causes them to move closer together.

18. Wind occurs when hot air rises in one place and cool air from another place moves in to take its place.

*Bonus Question: 10 points

19. Many factors are important in understanding the weather. See if you can list ten of them.

*Bonus questions will be included in the test at the end of the semester.

Choose answers from these terms.
All the terms may not be used and some may be used more than once:

hurricanes	thunderstorms	barometer	rain	thermometers
air	moist	dry	humidity	telephone
anemometer	sonar	lightweight	front	rotation
radar	telegraph	tornadoes	air pressure	wind

Fill in the Blank: Each question is worth 5 points.

1. When _____ air moves over an area, the _____ tends to be low, especially if the air is also warm.

2. A falling barometric reading usually indicates _____.

3. A barometer responds to the _____ _____ in the area.

4. A mass of _____ air results in a low-pressure area.

5. _____ measure the temperature of the air.

6. Hygrometers measure the _____ of the air.

7. Weather vanes and _____ can give specific information about wind direction and speed.

8. _____ and _____ devices can give detailed weather conditions as they develop.

9. Joseph Henry invented and used a _____ to communicate weather information to other cities.

10. A _____ is the leading edge of a large air mass.

11. _____ and _____ may result when the front edge of a mass of cold, dry air moves into a mass of warm, moist air.

12. _____ tend to originate in the tropical waters off the coast of Africa.

13. Besides differences in air pressure, temperature, and humidity, another factor that affects wind formation is the _____ of the earth.

14. _____ get energy from warm ocean water.

True / False: Each answer is worth 5 points.

15. Only increased snowfall is needed for an Ice Age to begin.

16. Volcanic activity after the Flood played a vital role in the formation of the Ice Age.

17. Geologists believe that sheets of ice covered a large portion of the United States during the Ice Age.

*Bonus Question: 10 points
Explain the difference between science and technology.

*Bonus questions will be included in the test at the end of the semester.

Choose answers from these terms.
All the terms may not be used and some may be used more than once:

front	newest	cooling	heavy	tornadoes
oldest	cirrus	bottom	sea	oceans
rivers	wind	plains	condensation	climate
sedimentary	decay	graveyards	estuaries	tilt
temperature	runoff	evolutionary	slope	evaporation
convection	prevailing winds	shelf	cumulus	weight
heating	cumulonimbus	rain	layers	rotation
lightweight	thunderstorms	tropical	groundwater	air pressure

Fill in the Blank: Each question is worth 2 points.

1. Most fossils are found in _____ rock.

2. Most fossils are the remains of _____-dwelling organisms.

3. In order to reach a more stable state, some radioactive atoms throw out little bits of nuclear particles, causing the original element to change into something else. This process is known as radioactive _____.

4. The principle of superposition states that the _____ layers, along with fossils and artifacts in them, are on the _____.

5. Fossil _____ occur when plants and animals from one environmental location may have been transported to another location.

6. The geologic column was established and agreed upon according to _____ ideas.

7. _____ contain fresh water; _____ contain a mixture of salt water and fresh water; _____ contain salt water.

8. Ocean zones differ primarily in the amount of water pressure, the food supply, the amount of sunlight, and the _____.

9. The continental _____ is the area bordering each continent, where the ocean floor gently slopes downward, as the ocean gradually gets deeper.

10. The deep ocean _____ are broad flat areas covered by thick layers of sediment.

11. The continental _____ is an area where the ocean floor suddenly becomes very steep, continuing downward until it reaches the flat ocean plains.

12. _____ currents can produce movements in any kind of fluid — air, water, and even magma.

13. Surface ocean currents are affected by several factors, but they tend to follow major _____ patterns.

14. Surface ocean currents all over the world are important in how they affect life and _____.

15. A drainage basin acts like a funnel, collecting all the _____ water within the area and channeling it toward the same point.

16. _____ and _____ are the two main processes in the water cycle.

17. Nimbus means _____.

18. Stratus means _____.

19. _____ are the tallest of all clouds.

20. The _____ of the earth is what causes the seasons.

21. A basic understanding of the weather can be obtained by understanding the effects of two main principles; the uneven _____ of the earth and the _____ of the air.

22. _____ _____ over the United States tend to blow from the west to the east.

23. A falling barometric reading usually indicates _____.

24. A barometer responds to the _____ _____ in the area.

25. A mass of _____ air results in a low-pressure area.

26. A _____ is the leading edge of a large air mass.

27. _____ and _____ may result when the front edge of a mass of cold, dry air moves into a mass of warm, moist air.

28. Besides differences in air pressure, temperature, and humidity, another factor that affects wind formation is the _____ of the earth.

29. _____ regions of the earth receive more direct rays from the sun than other regions.

True / False: Each answer is worth 2 points.

30. A paleontologist can conclude many things about an animal that left fossil footprints, including the kind of animal that left the footprint, how heavy it was, if it was limping, if it was running or walking, and the size of the animal.

31. There are very few places on earth where all the layers of the geologic column are found in the "official" order.

32. All ocean plants and animals are able to live anywhere in the ocean.

33. The total amount of liquid water on the earth has changed in the last 200 years.

34. Energy from the sun is what drives the water cycle.

35. When air molecules are heated, the increased heat energy causes them to move closer together.

36. Wind occurs when hot air rises in one place and cool air from another place moves in to take its place.

37. Volcanic activity after the Flood played a vital role in the formation of the Ice Age.

38. The rate at which radioactive decay occurs is not affected by temperature, pressure, or other chemicals.

39. When radioactive dating methods are used to date the age of certain rocks, they are always correct.

Short Answer Questions: Each of the following questions is worth 2 points:

40. Explain the important role that water plays in the formation of fossils.

41. The surface of the earth's continents consists mostly of sedimentary rock layers. These layers often extend over several states, sometimes over entire countries! How does water play a part in the formation of these layers?

42. Explain the water cycle.

43. How is the density of an object determined?

44. Explain the difference between science and technology.

Q	*Water & Weather* Concepts & Comprehension	Quiz 1 Level 2	Scope: Chapters 1–4	Total score: ____of 100	Name

Multiple Choice: Please select the best answer. Each question is worth 5 points:

1. Which of the following dinosaurs walked on two legs?
   a) Apatosaurus
   b) Allosaurus
   c) Triceratops

2. Which of the following dinosaurs walked on four legs?
   a) Triceratops
   b) T. rex
   c) Allosaurus

3. What are gastroliths?
   a) mammoths that have been found frozen with their fur, flesh, and bones intact
   b) rocks that seem to have been in the stomach region of some dinosaurs
   c) armor plates that covered the backs of dinosaurs

4. What do we call anything that fills in an impression made by a living plant or animal and hardens?
   a) a mold
   b) a trace fossil
   c) a cast

5. The Laetoli footprints were made in what material?
   a) volcanic ash
   b) wet sand
   c) Coconino Sandstone

6. The vast majority of fossils are organisms
   a) that walked on land
   b) that had a backbone
   c) that once lived in the sea

7. The vast majority of fossils that exist around the world
   a) are found on the ocean floor
   b) are found buried in ice
   c) are found in sedimentary rocks on dry land

8. When a dead plant or animal turns into a fossil, what role does dissolved minerals usually have in the process?
   a) the dissolved minerals form a protective coating around the organism
   b) chemicals in the dead plants and animals are replaced with the dissolved minerals
   c) dissolved minerals have nothing to do with fossil formation

9. Once a plant or animal turns into a hardened fossil by burial and mineral replacement, it will be surrounded by
   a) rock
   b) water
   c) carbon dioxide

10. Which processes can tear down layers of rocks?
   a) breakdown of carbon dioxide
   b) weathering and erosion
   c) rapid burial by layers of sedimentary deposits

Short Answer Questions: Each question is worth 10 points.

11. Read Job 41:1–34. On a piece of paper, draw each of the features listed, beginning with verse 12. Does your drawing look anything like an animal living today? Does it resemble a dinosaur?

12. Why do some scientists portray the Laetoli footprints as belonging to an ape-like animal, even though they clearly appear to be human footprints? What is wrong with this theory?

13. List four things that usually happen as once-living organisms turn to fossils.

   a.

   b.

   c.

   d.

14. Explain how a massive worldwide flood would have provided ideal and unrepeatable conditions for fossils to form.

15. What did scientist Mary Schweitzer find difficult to reconcile about the *T. Rex* bones found on an expedition to Montana a few years ago?

Multiple Choice: Please select the best answer. Each question is worth 5 points:

1. A radioactive element can change into a different element over time by the process of
    a) radioactive decay
    b) radioactive dating
    c) absolute dating

2. The half life of a radioactive element is used by many scientists as a tool for finding the ages of
    a) sedimentary rocks
    b) volcanic rocks
    c) igneous rocks

3. Determining the ratio of potassium to argon or uranium to lead are used to try to date
    a) old pieces of wood and cloth
    b) sedimentary layers
    c) volcanic rocks

4. Most fossils are found in
    a) sedimentary rocks
    b) igneous rocks
    c) volcanic rocks

5. Sedimentary layers harden by a process of
    a) erosion
    b) cementation and pressure
    c) uplifting

6. According to the principle of superposition, which layers are on the bottom of the geologic column?
    a) youngest
    b) oldest
    c) a mixture of the oldest and the youngest

7. Evidence of erosion between layers is called
    a) unconformities
    b) relative dating
    c) superposition

8. Intrusive rock formations may occur as _____ from deep below the ground squeezes through cracks in existing sedimentary rocks.
    a) water
    b) sediments
    c) liquid magma

9. Which of the following statements represents a creationist point of view?
    a) the oldest layers, along with fossils and artifacts in them, are on the bottom
    b) out-of-sequence layers may be explained by a giant folding of strata
    c) both a and b

10. Creation scientists agree with evolutionists that
    a) plants and animals evolved over millions of years
    b) millions of years passed between the formation of layers
    c) there were mass extinctions of plants and animals

Short Answer Questions: Each question is worth 10 points.

11. What is the difference between "absolute dating" and "relative dating"?

12. What is the definition of half life?

13. What does the geologic column supposedly show?

14. What is a creationist view of the geologic column?

15. What are index fossils?

Multiple Choice: Please select the best answer. Each question is worth 5 points:

1. When one solution floats on another solution, what does that tell about the densities of the two solutions?
   a) The density of the top solution is less than the density of the bottom solution
   b) The density of the bottom solution is less than the density of the top solution
   c) The density of both solutions is the same

2. In what ocean zone are most of the world's major fishing grounds found?
   a) intertidal zone
   b) deep zone
   c) neritic zone
   d) none of the above

3. What important food source grows in the neritic zone in huge numbers and is a main source of food for many kinds of fish?
   a) seaweed
   b) oysters
   c) plankton

4. The area bordering each continent, where the ocean floor gently slopes downward as the ocean gradually gets deeper is called the
   a) continental shelf
   b) continental rise
   c) abyssal plain

5. The area where the ocean floor suddenly becomes very steep, continuing downward until it reaches the bottom is called the
   a) oceanic trench
   b) continental slope
   c) continental shelf
   d) none of the above

6. What are the broad, flat areas covered by thick layers of sediment at the bottom of the ocean called?
   a) deep ocean plains
   b) magma
   c) oceanic trench

7. Small vehicles designed to withstand extreme pressure as they explore the deepest parts of the ocean are called
   a) submarines
   b) submersibles
   c) deep sea diving suits

8. The average depth of the ocean floor is
   a) 24 miles
   b) 2.4 miles
   c) 24 kilometers

9. The pressure on the ocean floor is how many times greater than air pressure at sea level?
   a) 40
   b) 400
   c) 4,000

Short Answer Questions: Each question is worth 5 points.

10. How is the density of an object determined?

11. What is the density of water?

12. What are estuaries?

13. What are three kinds of habitats found within the intertidal zone of the ocean?

14. What was discovered on the ocean floor that helps many living things to thrive in this harsh environment?

15. What are the factors that differentiate the various zones of the ocean?

16. Explain why scuba divers don't usually go past depths of 130 feet (40 meters).

17. Why are there no green plants 650 feet (200 meters) or more below the surface of the ocean?

Label the following diagram: Each answer is worth 1 point.

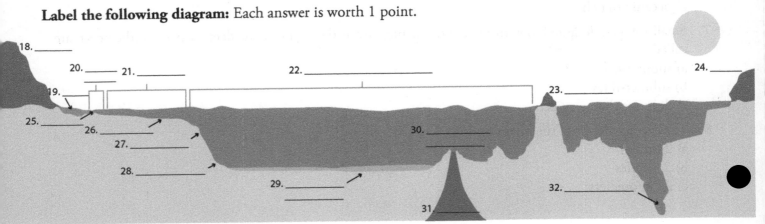

Multiple Choice: Please select the best answer. Each question is worth 5 points:

1. Surface ocean currents are affected by
   a) convection currents
   b) prevailing winds
   c) shape of the ocean floor
   d) all of the above

2. Surface ocean currents affect the
   a) top 100 feet of the ocean
   b) top 1,000 feet of the ocean
   c) top 10 feet of the ocean

3. _____ currents occur in the ocean when warm water flows toward the polar regions and cold water from the polar regions flows to the tropics.
   a) circular currents
   b) Gulf Stream currents
   c) convection currents

4. How does the Gulf Stream affect the weather on the eastern seaboard?
   a) the climate is much colder because the Gulf Stream keeps warm air from flowing north
   b) the climate is much warmer because the Gulf Stream carries warm air to the region
   c) the Gulf Stream does not affect the weather on the eastern seaboard

5. What did Magellan's voyage prove?
   a) that Christopher Columbus was wrong about ocean currents
   b) that the Gulf Stream would carry him from Spain to the new world
   c) that all the oceans of the world were connected
   d) all of the above

6. Which direction do surface ocean currents generally flow in the Northern Hemisphere?
   a) clockwise
   b) counterclockwise
   c) south to north

7. Which drainage basin is the largest in the United States?
   a) The Great Basin
   b) The Mississippi River Basin
   c) The Continental Divide Basin

8. What did Lewis and Clark discover about the Great Continental Divide?
   a) that water drains to the west on one side and to the east on the other
   b) that many rivers have their beginning at the Great Continental Divide
   c) both a and b are correct

9. Water from the Mississippi Drainage Basin may be carried to
   a) the Gulf of Mexico
   b) the Pacific Ocean
   c) the Great Salt Lake
   d) none of the above

10. When liquid water changes into water vapor, which process has occurred?
    a) evaporation
    b) precipitation
    c) condensation

11. When liquid water changes into water vapor, is energy absorbed or released by the water molecules?
    a) released
    b) absorbed
    c) energy is neither released or absorbed

12. Which statement is true about the water cycle?
    a) the processes of evaporation and precipitation balance each other
    b) the total amount of liquid water on the earth basically stays the same
    c) rainwater does not contain salt even though much of the rainwater comes from salty oceans
    d) all of the above

13. Which of these statements is true?
    a) exposing water to sunlight is a natural way of killing germs in the water
    b) adding chlorine to water will kill germs in the water
    c) filtration does not remove salts and chemicals that are dissolved in water
    d) all of the above

14. Most of the water on the earth is found as
    a) fresh water
    b) groundwater
    c) salt water

15. What is usually the first step in any man-made water purification process?
    a) killing germs
    b) identifying dissolved salts or chemicals in the water
    c) removing solid particles

Short Answer Questions: Each question is worth 5 points.

16. Why was knowledge of ocean currents important to early sailors?

17. Aside from discovering that a water route did not exist to the Pacific Ocean, describe another major impact of the Lewis and Clark expedition.

18. Explain why the Great Salt Lake is salty.

19. Describe the water cycle.

20. Where does most of the earth's drinking water come from?

Multiple Choice: Please select the best answer. Each question is worth 5 points:

1. High-level clouds form above
    a) 4 miles _____
    b) 8 miles
    c) 40 miles

2. Mid-level clouds typically appear between
    a) 1½ to 4 miles
    b) 4 miles to 8 miles
    c) 1 to 1 ½ miles

3. Low-level clouds usually lie below
    a) 4 miles
    b) 1 ½ miles
    c) 8 miles

4. Which is the tallest of all clouds?
    a) cumulus
    b) stratus
    c) cirrus
    d) cumulonimbus

5. What kind of weather can be produced from cumulonimbus clouds?
    a) rain
    b) thunderstorms
    c) tornadoes
    d) all of the above

6. For the three summer months at the North Pole, there is no night time. This is because
    a) the sun is very high on the horizon
    b) the sun is very low on the horizon
    c) the sun is not visible at all

7. The most direct rays of the sun strike an area
    a) in the morning
    b) in the evening
    c) at noontime

8. Seasons are caused by the fact that the earth is tilted on its axis at what angle?
    a) 22 ½
    b) 23
    c) 23 ½

9. The uneven heating of the earth produces
    a) wind
    b) direct rays from the sun
    c) slanted rays from the sun

10. Slight changes in the weight of the air are the result of
    a) the amount of water vapor in the air
    b) the temperature of the air
    c) the temperature of the land
    d) both a and b
    e) both b and c

11. In what general direction do the prevailing winds over the United States blow?
    a) east to west
    b) west to east
    c) north to south
    d) south to north

12. Would you expect the air in a weather "high" to weigh more or less than the air in a weather "low"?
    a) air in a high pressure area would weigh more
    b) air in a low pressure area would weigh more
    c) the air in both high and low pressure areas weigh the same

13. What kind of air would a "high" area be likely to contain?
    a) cold, dry air
    b) warm, humid air
    c) cold, humid air

14. Winds are deflected by
    a) the earth's spin
    b) the angle of the sun
    c) low areas of pressure
    d) high areas of pressure

15. Air molecules move faster and get farther apart when they are heated, causing what to happen?
    a) all the prevailing winds of the earth to blow from polar regions to the equator
    b) air gets heavier and hugs the ground
    c) air becomes lighter and rises

Short Answer Questions: Each question is worth 5 points.

16. Explain how a cloud can form from a mass of warm, humid air.

17. Suppose sunlight falls on a square meter of land at the equator and a square meter of land in an arctic region. Which square meter will absorb the most heat energy from the sun? Why is this?

18. Describe a cumulonimbus cloud.

19. What are the prefixes that are often added to the names of clouds to help describe them more accurately? What do they mean?

20. Explain the similarities and differences between dew and fog.

Multiple Choice: Please select the best answer. Each question is worth 5 points:

1. Which of these is the least useful in predicting next-day weather?
   a) thermometer
   b) rain guage
   c) hygrometer
   d) wind vane

2. A falling barometric reading usually indicates
   a) sunny weather
   b) coming rain
   c) high winds
   d) none of the above

3. Who persuaded the federal government to establish a national weather service?
   a) Steven O. Douglas
   b) Joseph Henry
   c) Benjamin Franklin
   d) Thomas Edison

4. The highs and lows on weather maps are surrounded by lines called isobars, which show
   a) air pressure
   b) wind direction
   c) wind speed
   d) temperature

5. In the United States, wind moves around a low pressure area in what direction?
   a) east to west
   b) west to east
   c) counterclockwise
   d) clockwise

6. When forecasters say a high-pressure area is moving toward your region, this usually means
   a) sunny weather
   b) cloudy weather
   c) tornadoes
   d) none of the above

7. What are some indicators that snow may be coming soon?
   a) falling barometric pressure
   b) approaching low-pressure system
   c) cumulonimbus clouds
   d) all of the above

8. Approximately how long did the Ice Age last?
   a) 70 years
   b) 700 years
   c) 7,000 years
   d) none of the above

9. Hurricanes do not have
   a) fronts
   b) counterclockwise winds
   c) a low pressure center
   d) warm, moist air

10. What results when there is a small difference in the temperature of two colliding air masses?
   a) strong winds
   b) gentle winds
   c) hurricanes
   d) typhoons

Short Answer Questions: Each question is worth 5 points.

11. Is stormy weather likely when a warm front moves into a mass of cold air?

12. What is a "front" as it pertains to weather?

13. What are three kinds of fronts?

14. Where do hurricanes typically originate?

15. In what direction do hurricanes travel?

16. Where do hurricanes get their energy?

17. What are two conditions that would be needed for an Ice Age to begin?

18. Explain why cold temperatures alone wouldn't cause an Ice Age.

19. What conditions might have occurred immediately after the Flood that could have caused the land to become cooler?

20. What conditions might have occurred immediately after the Flood that could have caused the oceans to become warmer? What could have caused more rain and snow to fall during this time?

Multiple Choice: Please select the best answer. Each question is worth 2 points.

1. What are gastroliths?
    a) mammoths that have been found frozen with their fur, flesh, and bones intact
    b) rocks that seem to have been in the stomach region of some dinosaurs
    c) armor plates that covered the backs of dinosaurs

2. What do we call anything that fills in an impression made by a living plant or animal and hardens?
    a) a mold
    b) a trace fossil
    c) a cast

3. The vast majority of fossils are organisms
    a) that walked on land
    b) that had a backbone
    c) that once lived in the sea

4. The vast majority of fossils that exist around the world
    a) are found on the ocean floor
    b) are found buried in ice
    c) are found in sedimentary rocks on dry land

5. When a dead plant or animal turns into a fossil, what role does dissolved minerals usually have in the process?
    a) the dissolved minerals form a protective coating around the organism
    b) chemicals in the dead plants and animals are replaced with the dissolved minerals
    c) dissolved minerals have nothing to do with fossil formation

6. Which processes can tear down layers of rocks?
    a) breakdown of carbon dioxide
    b) weathering and erosion
    c) rapid burial by layers of sedimentary deposits

7. A radioactive element can change into a different element over time by the process of
    a) radioactive decay
    b) radioactive dating
    c) absolute dating

8. The half life of a radioactive element is used by many scientists as a tool for finding the ages of
    a) sedimentary rocks
    b) volcanic rocks
    c) igneous rocks

9. According to the principle of superposition, which layers are on the bottom of the geologic column?
    a) youngest
    b) oldest
    c) a mixture of the oldest and the youngest

10. Which of the following statements represents a creationist point of view?
    a) the oldest layers, along with fossils and artifacts in them, are on the bottom
    b) out of sequence layers may be explained by a giant folding of strata
    c) both a and b

11. Creation scientists agree with evolutionists that
    a) plants and animals evolved over millions of years
    b) millions of years passed between the formation of layers
    c) there were mass extinctions of plants and animals

12. When one solution floats on another solution, what does that tell about the densities of the two solutions?
    a) The density of the top solution is less than the density of the bottom solution
    b) The density of the bottom solution is less than the density of the top solution
    c) The density of both solutions is the same

13. In what ocean zone are most of the world's major fishing grounds found?
    a) intertidal zone
    b) deep zone
    c) neritic zone
    d) none of the above

14. The area bordering each continent, where the ocean floor gently slopes downward as the ocean gradually gets deeper is called the
    a) continental shelf
    b) continental rise
    c) abyssal plain

15. The average depth of the ocean floor is
    a) 24 miles
    b) 2.4 miles
    c) 24 kilometers

16. The pressure on the ocean floor is how many times greater than air pressure at sea level?
    a) 40
    b) 400
    c) 4,000

17. Surface ocean currents are affected by
    a) convection currents
    b) prevailing winds
    c) shape of the ocean floor
    d) all of the above

18. _____ currents occur in the ocean when warm water flows toward the polar regions and cold water from the polar regions flows to the tropics.
    a) circular currents
    b) Gulf Stream currents
    c) convection currents

19. Which drainage basin is the largest in the United States?
    a) The Great Basin
    b) The Mississippi River Basin
    c) The Continental Divide Basin

20. When liquid water changes into water vapor, which process has occurred?
    a) evaporation
    b) precipitation
    c) condensation

21. Most of the water on the earth is found as
    a) fresh water
    b) groundwater
    c) salt water

22. For the three summer months at the North Pole, there is no nighttime. This is because
    a) the sun is very high on the horizon
    b) the sun is very low on the horizon
    c) the sun is not visible at all

23. Seasons are caused by the fact that the earth is tilted on its axis at what angle?
    a) 22 ½
    b) 23
    c) 23 ½

24. Slight changes in the weight of the air are the result of
    a) the amount of water vapor in the air
    b) the temperature of the air
    c) the temperature of the land
    d) both a and b
    e) both b and c

25. Would you expect the air in a weather "high" to weigh more or less than the air in a weather "low"?
    a) air in a high pressure area would weigh more
    b) air in a low pressure area would weigh more
    c) the air in both high and low pressure areas weigh the same

26. The highs and lows on weather maps are surrounded by lines called isobars, which show
    a) air pressure
    b) wind direction
    c) wind speed
    d) temperature

27. What are some indicators that snow may be coming soon?
    a) falling barometric pressure
    b) approaching low pressure system
    c) cumulonimbus clouds
    d) all of the above

28. Approximately how long did the Ice Age last?
    a) 70 years
    b) 700 years
    c) 7,000 years
    d) none of the above

29. How does the Gulf Stream affect the weather on the eastern seaboard?
    a) the climate is much colder because the Gulf Stream keeps warm air from flowing north
    b) the climate is much warmer because the Gulf Stream carries warm air to the region
    c) the Gulf Stream does not affect the weather on the eastern seaboard

30. When liquid water changes into water vapor, is energy absorbed or released by the water molecules?
   a) released
   b) absorbed
   c) energy is neither released nor absorbed

Short Answer Questions: Each of the following questions is worth 2 points:

31. Explain how a massive worldwide flood would have provided ideal and unrepeatable conditions for fossils to form.

32. What does the geologic column supposedly show?

33. What is a creationist view of the geologic column?

34. What are the factors that differentiate the various zones of the ocean?

35. Why are there no green plants below 650 feet (200 meters) of the surface of the ocean?

36. Describe the water cycle.

37. Explain how a cloud can form from a mass of warm, humid air.

38. Explain why cold temperatures alone wouldn't cause an Ice Age.

Bonus Question
Label the following diagram: Each answer is worth 1 point.

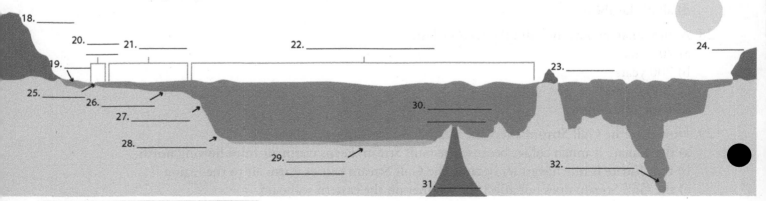

Quizzes and Test

for Use with

The Universe

Testing:

This series is appropriate for both upper elementary and junior high students. Because of this, we have included quizzes and tests in two different levels which you can choose from based on your child's abilities and understanding of the concepts in the course.

Level 1: suggested for younger ages or those who struggle with application of the concepts beyond just definitions and basic concepts

Level 2: suggested for older ages or those who can both grasp the scientific concepts and apply them consistently

Choose answers from these terms.
All the terms may not be used and some may be used more than once:

constellations	supernova	nuclear fusion	hydrogen nuclei
variable	ultraviolet waves	infrared waves	sundial
magnetic field	stable	helium nuclei	spectroscope
black light	spectrum	electromagnetic	nebula

Fill in the Blank: Each question is worth 5 points.

1. When light rays pass through a prism, the visible rays separate into a _____ of colors.

2. The entire range of waves that come from the sun are known as _____ waves.

3. Sunscreen filters out _____ _____.
   _____ _____ are detected by night-vision goggles.

4. A _____ provided an accurate means of telling time before mechanical clocks were invented.

5. Electromagnetic waves that leave the sun and head toward the earth are deflected by the earth's _____ _____.

6. Our sun is a _____ star. Some stars are _____ stars.

7. A _____ is an instrument used by scientists to produce a spectrum of visible light from the sun and other stars.

8. _____ are groups of stars that have been given names.

9. Occasionally, a star may explode as a _____, leaving behind a _____.

Short Answer: Each question is worth 5 points

10. Electromagnetic waves travel through space at what speed?

11. Which are more dangerous — very short electromagnetic waves or very long electromagnetic waves?

12. In what city is there a special telescope that is used to pinpoint noon when the sun is exactly overhead, and is used to set all clocks on earth?

13. How many constellations are recognized by the International Astronomical Union?

14. Name three constellations mentioned in the Book of Job.

    a.

    b.

    c.

15. Name at least three ways stars can differ from one another.

    a.

    b.

    c.

True / False Questions: Each question is worth 2 points:

16. All electromagnetic waves travel at the same speed.

17. All electromagnetic waves have the same amount of energy and wavelengths.

18. Energy released by nuclear fusion emerges as light, heat, and other cosmic radiations.

19. The sun turns slowly on its axis. According to the nebula theory, it should be turning much slower than it is.

20. Isaac Newton said of the universe, "This most beautiful system could only proceed from the counsel and dominion of an intelligent and powerful being."

Bonus question: Worth 10 points:

21. Briefly explain the two main explanations for how things in the universe came to exist.

*Bonus questions will be included in the test at the end of the semester.

Fill in the Blank Questions: (2 Points Each)

Choose answers from these terms.
All the terms may not be used and some may be used more than once:

Newton	supernatural	larger	helium
frozen	meteorites	gravitational	Kepler
spectrum	rings	Jovian	predictable
TNO	carbon dioxide	elliptical	Copernicus
reaction	eclipse	natural world	nitrogen
nebula theory	galaxy	hydrogen	spectrum
0xygen	electromagnetic	friction	
solar system	Galileo	magnetic field	

1. The _____ region is beyond Neptune.

2. The _____ _____ is made up of the sun and the planets that orbit the sun, along with everything else held in place by the sun's gravity.

3. A _____ is made up of millions or billions of stars that are bound together by gravitational attraction.

4. The universe is a term used to include everything that exists in the _____ _____.

5. The _____ _____ proposes that, following the big bang, there were swirling clouds throughout space. These clouds eventually formed swirling eddies of gas that condensed into stars and planets and all the other objects in space.

6. Science can only investigate the way in which the universe operates today. It has no way of explaining _____ events.

7. Kepler is most remembered for discovering that planets move around the sun in _____ _____ orbits.

8. A falling object is pulled down by a _____ force.

9. Air _____ exerts an upward force on a falling object.

10. _____ built a telescope that enabled him to see some of the moons around Jupiter.

11. Rockets operate on the basis of _____'s third law of motion that says "For every action, there is an equal and opposite _____."

12. The gravitational pull of the large gas planets in our solar system helps to protect the earth from being hit by _____.

13. The earth maintains a balance of _____ and _____ _____ at all times.

14. The moon is exactly the right size to just cover the sun during an _____ of the sun.

15. _____ in the atmosphere causes most meteorites that come close to the earth to burn up before they hit the earth.

16. Jupiter, Saturn, Uranus, and Neptune, are known as the _____ planets.

17. The two main elements that make up the gas planets are _____ and _____.

18. All of the gas planets have multiple _____ and moons around them.

19. The gas planets are much _____ than the rocky planets.

20. Outer space is very cold. If the billions of years theory were correct we would expect most celestial bodies without atmospheres to become _____ solid.

21. When light rays pass through a prism, the visible rays separate into a _____ of colors.

22. The entire range of waves that come from the sun are known as _____ waves.

23. Electromagnetic waves that leave the sun and head toward the earth are deflected by the earth's _____ _____.

True / False Questions: Each question is worth 2 points:

24. The Milky Way is the name given to a cluster of a few dozen galaxies that contains our solar system.

25. Before the time of Hubble, scientists believed all the stars in the universe were in the same galaxy.

26. Galileo noted that as long as the force of gravity kept being applied to a rolling ball on a ramp, the ball continued to accelerate.

27. Galileo found a mathematical formula that could be used to calculate the gravitational attraction anywhere in the universe.

28. When spacecraft re-enter the earth's atmosphere, the friction between the air and the spacecraft creates tremendous heat.

29. The earth's atmosphere is made of about 20 percent oxygen and about 80 percent nitrogen.

30. One side of the moon is never visible from the earth.

31. All electromagnetic waves travel at the same speed.

32. All electromagnetic waves have the same amount of energy and wavelengths.

33. Not all of the rocky planets have an atmosphere.

Short Answer: Each question is worth 2 points:

34. What does AU stand for?

35. What is a light year?

36. What did Einstein name his two theories of relativity?

37. What is the basic difference between Einstein's two theories of relativity?

38. Does the earth get warmer when it is closer to the sun?

39. Why would a feather and a hammer, dropped on the moon at the same time, from the same height, hit the ground at the same time?

40. Why were people never able to attach artificial wings to their bodies and fly?

41. On the earth, how does the atmosphere affect the temperature at night when there is no sunlight on the earth?

42. List three of the several conditions that are "just right" on earth or in our solar system, such that life would be difficult if they were different.

    a.

    b.

    c.

43. What does it mean to say that the angular size of the moon and the angular size of the sun are nearly equal?

44. Briefly tell what happened during the Enlightenment Period of history.

45. Name three constellations mentioned in the Book of Job.

    a.

    b.

    c.

46. Name at least three ways stars can differ from one another.

    a.

    b.

    c.

Multiple Choice: Please select the best answer. Each question is worth 5 points:

1. An Astronomical Unit (AU) is equal to
   a) the distance from the earth to the moon
   b) the distance from the moon to the sun
   c) the distance from the earth to the sun
   d) none of the above

2. Which of the following is not a rocky planet?
   a) Mars
   b) Saturn
   c) Venus
   d) all of the above

3. The distance from the sun to the earth is equal to
   a) 93,000,000,000 miles
   b) 93,000,000 miles
   c) 93,000 miles
   d) none of the above

4. How many miles equal a light year?
   a) 6 trillion
   b) 60 trillion
   c) 60 billion
   d) none of the above

5. The nearest star to our sun is
   a) Tritan
   b) Titan
   c) Pleiades
   d) Alpha Centauri

6. What did Einstein name his two theories of relativity?
   a) special theory of relativity
   b) revised theory of relativity
   c) general theory of relativity
   d) both b and c above
   e) both a and c above

7. Einstein found that
   a) velocity and gravity can affect time
   b) light always travels in a straight line
   c) gravitational force has no impact on the laws of motion
   d) none of the above

8. What makes it difficult to detect movement in objects that are extremely far away?
   a) the nebula theory
   b) the parallax effect
   c) the direction the object is rotating
   d) all of the above

9. Which color in the visible spectrum has the longest wavelength?
   a) red
   b) violet
   c) blue
   d) yellow

10. Who discovered that the universe contained more than one galaxy?
   a) Galileo
   b) Copernicus
   c) Hubble
   d) Kepler

Short Answer Questions: Each question is worth 5 points.

11. What is the solar system?

12. What is a galaxy?

13. What is the universe?

14. What is the Milky Way?

15. What is the Local Group?

16. According to Einstein's special theory of relativity, what is the fastest speed that anything can reach?

17. What is the basic difference between Einstein's two theories of relativity?

18. Under what conditions might Newton's laws of motion not apply?

19. The "big-bang" theory is based on what main piece of evidence?

20. What evidence did Edwin Hubble discover that caused him to conclude that galaxies are moving and getting farther away from the earth?

Multiple Choice: Please select the best answer. Each question is worth 5 points:

1. Who determined that planets move in elliptical orbits rather than in round circles?
   a) Kepler
   b) Copernicus
   c) Galileo
   d) Hubble

2. What exerts an upward force on a falling object?
   a) friction
   b) gravity
   c) inertia
   d) none of the above

3. Who was one of the first scientists to insist that scientific ideas and explanations need to be tested?
   a) Kepler
   b) Copernicus
   c) Aristotle
   d) Galileo

4. Around what time period did scientists begin to use telescopes to study planets, moons, and other objects in space?
   a) early 1500s
   b) early 1600s
   c) early 1700s
   d) early 1800s

5. Moving objects tend to stay in motion and objects that are not moving tend to keep on not moving. This property is called
   a) gravity
   b) friction
   c) inertia
   d) none of the above

6. What kind of lens is a magnifying glass?
   a) concave
   b) mirrored
   c) refractive
   d) convex

7. What is another word for bending light?
   a) reflection
   b) magnification
   c) spectrolysis
   d) refraction

8. What famous scientist built a telescope that enabled him to see some of the moons around Jupiter?
    a) Kepler
    b) Hubble
    c) Copernicus
    d) Galileo

9. The outer planets travel around the sun in more
    a) circular orbits
    b) elliptical orbits
    c) unpredictable orbits
    d) none of the above

10. Aristotle based many of his conclusions about science on
    a) logic
    b) instinct
    c) scientific tests
    d) none of the above

Short Answer Questions: Each question is worth 5 points.

11. What causes the different seasons of the earth?

12. The amount of gravitational attraction that exists between objects depends on what two things?

13. Explain why a heavy rock and a lightweight rock will hit the ground at the same time if they are dropped from the same height.

14. What is the difference in how a refracting telescope and a microscope are made?

15. Does the earth get warmer when it is closer to the sun?

Multiple Choice: Please select the best answer. Each question is worth 5 points:

1. What did the Wright brothers have to spend months doing before they were able to actually fly the first airplane?
   a) testing
   b) rebuilding
   c) designing
   d) all of the above

2. Who discovered the principle of floating in air?
   a) Archimedes
   b) Aristotle
   c) Galileo
   d) Copernicus

3. What was the first gas to replace the hot air that allowed people to travel long distances through the air?
   a) helium
   b) propane
   c) hydrogen
   d) none of the above

4. What happens when a spacecraft re-enters the earth's atmosphere?
   a) friction between the air and the spacecraft creates heat
   b) it must be protected by a heat shield
   c) it is traveling at a high rate of speed
   d) all of the above

5. How much of the air we breathe is made up of oxygen?
   a) 80%
   b) 100%
   c) 20%
   d) none of the above

6. What is the average air pressure at sea level?
   a) 14.7 lbs. per inch
   b) 1.47 lbs. per inch
   c) 147 lbs. per inch
   d) none of the above

7. From what danger does the earth's magnetic field protect the earth?
   a) high-energy particles and cosmic rays
   b) cosmic rays and meteors
   c) high-energy particles and solar flares
   d) none of the above

8. Which layer of the atmosphere is composed mainly of small amounts of hydrogen and helium, the lightest gases on earth?
   a) troposphere
   b) exosphere
   c) thermosphere
   d) stratosphere

9. Which layer of the atmosphere contains the air that we breathe?
   a) troposphere
   b) exosphere
   c) thermosphere
   d) stratosphere

10. Which layer of the atmosphere contains a layer of ozone?
    a) troposphere
    b) exosphere
    c) thermosphere
    d) stratosphere

11. Which layer of the atmosphere connects with outer space?
    a) troposphere
    b) exosphere
    c) thermosphere
    d) stratosphere

12. Which of these conditions are "just right" to make life on earth possible?
    a) the right amount of gravitational pull to hold the atmosphere in place
    b) enough oxygen to live but not so much that fires would be uncontrollable
    c) air pressure that is enough to balance internal pressures
    d) all of the above

Short Answer Questions: Each question is worth 5 points.

13. What famous airship burned and crashed, putting an end to travel by air ships?

14. What is Newton's third law of motion?

15. How do plants and animals maintain a balance of the amount of carbon dioxide and oxygen in the air?

16. Name at least three positive uses for rockets.

17. Name the five layers of the atmosphere in order, starting with the lowest level.

18. Why do planets and moons with no atmosphere or very thin air not have liquid water on them?

19. Why is the earth not covered with craters like the moon?

20. Explain why Venus does not have seasons.

Multiple Choice: Please select the best answer. Each question is worth 5 points:

1. Every planet in the solar system except this one has been visited by spacecraft.
   a) Mercury
   b) Neptune
   c) Saturn
   d) all of the above

2. Which planet has a thick atmosphere of carbon dioxide and clouds of sulfuric acid?
   a) Venus
   b) Saturn
   c) Uranus
   d) Mars

3. Besides the moon, what is one of the brightest lights in the night sky?
   a) Mars
   b) Venus
   c) Mercury
   d) none of the above

4. Which planet turns on its axis opposite to all the other planets?
   a) Jupiter
   b) Saturn
   c) Venus
   d) Uranus

5. Which of these planets have an atmosphere?
   a) Mercury
   b) Venus
   c) Mars
   d) all of the above

6. On which of the rocky planets is the sunset in the east?
   a) Earth
   b) Mercury
   c) Mars
   d) Venus

7. Which is the largest Jovian planet?
   a) Saturn
   b) Jupiter
   c) Uranus
   d) Neptune

8. Which Jovian planet has the most moons?
   a) Saturn
   b) Uranus
   c) Jupiter
   d) Neptune

9. Which Jovian planet has the longest day?
  a) Saturn
  b) Uranus
  c) Jupiter
  d) Neptune

10. Which of these planets has rings around it?
  a) Saturn
  b) Jupiter
  c) Uranus
  d) all of the above

Short Answer Questions: Each question is worth 5 points.

11. Why is only one side of the moon seen from the earth?

12. What determines which part of the moon is visible at night?

13. How does the earth's gravitational pull affect the moon?

14. What does it mean to say the angular size of the moon and the angular size of the sun are nearly equal?

15. Briefly explain what happened during the Enlightenment Period of history.

Multiple Choice: Please select the best answer. Each question is worth 5 points:

1. How do electromagnetic waves differ from each other?
   a) wavelengths
   b) frequencies
   c) amount of energy
   d) all of the above

2. Which of these electromagnetic waves are longer than visible light?
   a) radar
   b) TV
   c) AM radio
   d) all of the above

3. Do very short electromagnetic waves vibrate faster or slower than long waves?
   a) they vibrate faster
   b) they vibrate slower
   c) they vibrate at the same speed

4. What percentage of the solar system does the sun's mass make up?
   a) 80 %
   b) 90%
   c) 95%
   d) 99.5 %

5. Electromagnetic waves that leave the sun and head toward the earth are deflected by what?
   a) solar flares
   b) the earth's magnetic field
   c) the ozone
   d) all of the above

6. How many constellations are recognized today by the International Astronomical Union?
   a) 85
   b) 90
   c) 88
   d) 78

7. What is the best explanation for why a different zodiac constellation appears each month?
   a) because the earth moves around the sun
   b) because the moon moves around the earth
   c) because the stars move around the sun

8. What is another name for an exploding star?
   a) nebula
   b) comet
   c) supernova
   d) none of the above

9. Which of these have been found in space?
    a) tiny atomic particles
    b) electromagnetic waves
    c) dust
    d) all of the above

10. Which of these would have the hottest temperature?
    a) red star
    b) blue star
    c) yellow star
    d) they all have roughly the same temperature

Short Answer Questions: Each question is worth 5 points.

11. How do sound waves differ from light waves and other kinds of electromagnetic waves?

12. Briefly describe the two main explanations for how things in the universe came to exist.

13. What kind of electromagnetic wave is detected by night-vision goggles?

14. Electromagnetic waves travel through space at what speed?

15. What do scientists believe is the source of the energy that comes from the sun?

Multiple Choice: (2 Points each)

1. The distance from the sun to the earth is equal to
   a) 93,000,000,000 miles
   b) 93,000,000 miles
   c) 93,000 miles
   d) none of the above

2. How many miles equal a light year?
   a) 6 trillion
   b) 60 trillion
   c) 60 billion
   d) none of the above

3. The nearest star to our sun is
   a) Tritan
   b) Titan
   c) Pleiades
   d) Alpha Centauri

4. Einstein found that
   a) velocity and gravity can affect time
   b) light always travels in a straight line
   c) gravitational force has no impact on the laws of motion
   d) none of the above

5. What makes it difficult to detect movement in objects that are extremely far away?
   a) the nebula theory
   b) the parallax effect
   c) the direction the object is rotating
   d) all of the above

6. Who discovered that the universe contained more than one galaxy?
   a) Galileo
   b) Copernicus
   c) Hubble
   d) Kepler

7. What exerts an upward force on a falling object?
   a) friction
   b) gravity
   c) inertia
   d) none of the above

8. Who was one of the first scientists to insist that scientific ideas and explanations need to be tested?
   a) Kepler
   b) Copernicus
   c) Aristotle
   d) Galileo

9. Moving objects tend to stay in motion and objects that are not moving tend to keep on not moving. This property is called
   a) gravity
   b) friction
   c) inertia
   d) none of the above

10. What kind of lens is a magnifying glass?
   a) concave
   b) mirrored
   c) refractive
   d) convex

11. What is another word for bending light?
   a) reflection
   b) magnification
   c) spectrolysis
   d) refraction

12. The outer planets travel around the sun in more
   a) circular orbits
   b) elliptical orbits
   c) unpredictable orbits
   d) none of the above

13. What was the first gas to replace the hot air that allowed people to travel long distances through the air?
   a) helium
   b) propane
   c) hydrogen
   d) none of the above

14. How much of the air we breathe is made up of oxygen?
   a) 80%
   b) 100%
   c) 20%
   d) none of the above

15. What is the average air pressure at sea level?
   a) 14.7 lbs. per inch
   b) 1.47 lbs. per inch
   c) 147 lbs. per inch
   d) none of the above

16. From what danger does the earth's magnetic field protect the earth?
   a) high energy particles and cosmic rays
   b) cosmic rays and meteors
   c) high energy particles and solar flares
   d) none of the above

17. Which layer of the atmosphere contains the air that we breathe?
   a) troposphere
   b) exosphere
   c) thermosphere
   d) stratosphere

18. Which layer of the atmosphere contains a layer of ozone?
    a) troposphere
    b) exosphere
    c) thermosphere
    d) stratosphere

19. Besides the moon, what is one of the brightest lights in the night sky?
    a) Mars
    b) Venus
    c) Mercury
    d) none of the above

20. Which planet turns on its axis opposite to all the other planets?
    a) Jupiter
    b) Saturn
    c) Venus
    d) Uranus

21. Which of these planets have an atmosphere?
    a) Mercury
    b) Venus
    c) Mars
    d) all of the above

22. Which is the largest Jovian planet?
    a) Saturn
    b) Jupiter
    c) Uranus
    d) Neptune

23. Which Jovian planet has the most moons?
    a) Saturn
    b) Uranus
    c) Jupiter
    d) Neptune

24. Which of these planets has rings around it?
    a) Saturn
    b) Jupiter
    c) Uranus
    d) all of the above

25. How do electromagnetic waves differ from each other?
    a) wavelengths
    b) frequencies
    c) amount of energy
    d) all of the above

26. Do very short electromagnetic waves vibrate faster or slower than long waves?
    a) they vibrate faster
    b) they vibrate slower
    c) they vibrate at the same speed

27. What percentage of the solar system does the sun's mass make up?
    a) 80 %
    b) 90%
    c) 95%
    d) 99.5 %

28. Electromagnetic waves that leave the sun and head toward the earth are deflected by what?
    a) solar flares
    b) the earth's magnetic field
    c) the ozone
    d) all of the above

29. How may constellations are recognized today by the International Astronomical Union?
    a) 85
    b) 90
    c) 88
    d) 78

30. Which of these would have the hottest temperature?
    a) red star
    b) blue star
    c) yellow star
    d) they all have roughly the same temperature

Short Answer Questions: Each question is worth 5 points.

31. According to Einstein's special theory of relativity, what is the fastest speed that anything can reach?

32. What is the basic difference between Einstein's two theories of relativity?

33. What evidence did Edwin Hubble discover that caused him to conclude that galaxies are moving and getting farther away from the earth?

34. The amount of gravitational attraction that exists between objects depends on what two things?

35. Explain why a heavy rock and a lightweight rock will hit the ground at the same time if they are dropped from the same height.

36. Why do planets and moons with no atmosphere or very thin air not have liquid water on them?

37. Briefly explain what happened during the Enlightenment Period of history.

38. What do scientists believe is the source of the energy that comes from the sun?

Quiz and Test Answers

for Use with

Science Starters: Elementary General Science & Astronomy

Water & Weather
Quiz Answer Keys
Level 1 & 2

Quiz 1 Level 1, chapters 1-4

1. rocks
2. dinosaurs, eggs, offspring
3. reptiles
4. mammals
5. paleontologists, distance
6. sedimentary
7. sea
8. location, environment
9. An animal that was covered by a large amount of sediment
10. No
11. Fossils are common all over the earth.
12. No
13. True
14. False
15. False
16. True
17. When an organism changes into a fossil, the minerals in the dead organism are replaced by minerals in the surrounding mud or water and cemented together.

Quiz 2 Level 1, chapters 5-6

1. decay
2. daughter, half life
3. sedimentary
4. index fossils
5. oldest, bottom
6. erosion
7. magma
8. worldwide flood
9. graveyards
10. evolutionary
11. No
12. True
13. False
14. True
15. False
16. True
17. True
18. True
19. Sediments were laid down by water in flat layers that eventually harden into rock. The shape of the layers can also be altered by water. An example of this would be heavy rains eroding the layers to form canyons.

Quiz 3 Level 1, chapters 7-9

1. rivers, estuaries, oceans
2. salt lake
3. sediment/delta
4. minerals
5. brackish
6. salinity
7. water
8. zones
9. temperature
10. weight
11. greater
12. submersibles
13. shelf
14. plains
15. slope
16. False
17. True
18. True
19. density = mass/volume

Quiz 4 Level 1, chapters 10-13

1. streams
2. convection
3. wind
4. clockwise
5. colder
6. climate
7. watersheds
8. runoff
9. Mississippi River
10. Continental Divide
11. prevailing winds
12. water cycle
13. evaporation, condensation
14. groundwater
15. germs
16. False
17. False
18. True
19. True
20. Water molecules evaporate from the earth after being warmed by the sun. As the water vapor cools at higher altitudes, the vapor condenses into liquid water, which may fall to the earth in the form of rain or snow.

Quiz 5 Level 1, chapters 14-16

1. cirrus
2. cumulus
3. fog
4. humid
5. rain
6. layers
7. cumulonimbus
8. dew
9. tilt
10. slanted
11. revolution
12. tropical
13. heating, weight
14. prevailing winds
15. satellites
16. concentrated, more
17. False
18. True
19. Clouds, the angles of the sun's rays, temperature of the air, temperatures of the land and oceans, weight of the air, high and low air-pressure systems, the makeup of large moving and stationary air masses, the moving of different kinds of air fronts, water vapor in the air, wind speed, wind direction, prevailing winds

Quiz 6 Level 1, chapters 17-20

1. moist, barometer
2. rain
3. air pressure
4. lightweight
5. thermometers
6. humidity
7. anemometers
8. radar, sonar
9. telegraph
10. front
11. thunderstorms, tornadoes
12. hurricanes
13. rotation
14. hurricanes
15. False
16. True
17. True
18. Pure science provides explanations for things that have been observed in nature, while technology is about making things that have a useful purpose.

Quiz 1 Level 2, chapters 1-4

1. b
2. a
3. b
4. c
5. a
6. c
7. c
8. b
9. a
10. b
11. Answers will vary.
12. According to radiometric dating tests, the volcanic ash in which the footprints were found was about 3.5 million years old. They assumed there were no humans around 3.5 million years ago. Radiometric dating tests of volcanic rocks are not always accurate.
13. (a) The dead remains are quickly covered up. This prevents them from being eaten. (b) The oxygen supply is cut off, which will cause most decay to stop. (c) There is pressure (compaction) on the plant or animal remains. (d) There is a sufficient amount of fossilizing, dissolved minerals around the remains.
14. Rapid burial by layers of sedimentary deposits would have kept remains of plants and animals from being eaten, and the sediment would have quickly cut off the oxygen supply. Minerals necessary for fossilization would have been dissolved in the water. The weight of the water and deep layers of sediments would have resulted in pressure on the remains.
15. Soft tissue, vessels, and small red structures resembling red blood cells were found inside the bones — yet the bones were estimated to be 65 million years old.

Quiz 2 Level 2, chapters 5-6

1. a
2. c
3. c
4. a
5. b
6. b
7. a
8. c
9. c
10. c
11. In absolute dating, the age of an object has been determined by using a method that gives a specific date — regardless of whether the date is correct or not, such as counting the rings on a tree. Relative dating methods try to determine if one rock is older than another rock, regardless of whether the age difference is days or years.

12. The length of time it takes for one-half of the original radioactive element to change into daughter elements.
13. The progression of the evolution of all living things, along with the ages in which different groups evolved
14. The progression of events that occurred during the Flood
15. Fossils of certain plants or animals that lived and flourished for a long time in many places around the world, then died out when there were circumstances that led to their extinction

Quiz 3 Level 2, chapters 7-9

1. a
2. c
3. c
4. a
5. b
6. a
7. b
8. a
9. b
10. By dividing the weight (or mass) of the object by its volume (density=mass/volume)
11. 1 gram per milliliter (1 g/mL)
12. Estuaries are inlets or bays along the coast where fresh water from rivers mixes with salt water from the ocean.
13. Rocky shores with tidal pools, coastal wetlands, estuaries
14. Thermal vents or hot water vents
15. Temperature, amount of sunlight, amount of water pressure, food supply
16. Beyond that depth, the increased water pressure makes it difficult for a diver to live and function well.
17. Sunlight can only penetrate about 650 feet.
18.–32. See chart on next page

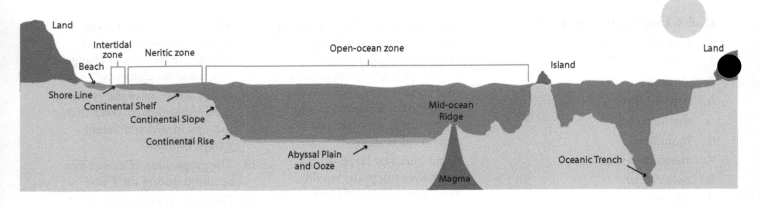

Land

Intertidal zone

Neritic zone

Open-ocean zone

Island

Land

Beach

Shore Line

Continental Shelf

Continental Slope

Continental Rise

Abyssal Plain and Ooze

Mid-ocean Ridge

Oceanic Trench

Magma

Quiz 4 Level 2, chapters 10-13

1. d
2. b
3. c
4. b
5. c
6. a
7. b
8. c
9. a
10. a
11. b
12. d
13. d
14. c
15. c
16. Ships could be carried across the ocean in a predictable pattern.
17. It marked the beginning of a major westward expansion of the United States.
18. There is no drainage system for the water. As the water in the lake is warmed by the sun it evaporates leaving behind the chemicals in the water.
19. Water molecules evaporate from the earth after being warmed by the sun. As the water vapor cools at higher altitudes, the vapor condenses into liquid water which may fall to the earth in the form of rain or snow.
20. Lakes, rivers, and groundwater

Quiz 5 Level 2, chapters 14-16

1. a

2. a
3. b
4. d
5. d
6. b
7. c
8. c
9. a
10. d
11. b
12. a
13. a
14. a
15. c
16. The warm, humid air tends to rise in the air. The higher it rises, the cooler the air becomes. Eventually the moisture in the air condenses into tiny droplets and forms a cloud.
17. A square meter of land at the equator will absorb more heat energy than a square meter of land in an arctic region. This is because the sun's rays hit the earth more directly at the equator than it does in arctic regions.
18. They are the tallest of all clouds. Strong convection updrafts can cause them to form towers that extend as high as 7 or 8 miles. They can produce rain, thunderstorms, and tornadoes. They tend to have large anvil-shaped tops.
19. "Alto" means middle, "Fracto" means broken, "Nimbo" means rain

20. Dew is similar to fog in that cold air causes water vapor to condense into tiny droplets. Dew is different from fog in that the droplets condense on the ground instead of floating in the air

Quiz 6 Level 2, chapters 17-20

1. b
2. b
3. b
4. a
5. c
6. a
7. d
8. b
9. a
10. b
11. No
12. A front is the leading edge of a large air mass.
13. Cold front, warm front, stationary front
14. In tropical waters near West Africa
15. Hurricanes move across the Atlantic Ocean from the east to the west until they start to turn north.
16. Warm ocean water
17. Increased snowfall and cooler summers
18. As the air becomes colder, it cannot hold much water vapor. Since very cold air is also dry, it will not produce much snow. Increased snowfall occurs when the air is not too cold. Snow and

ice accumulate when summer temperatures are cool enough that much of the snow doesn't melt. Then the next year's snows will add another layer of ice to what is already on the ground. Also, any liquid water on the ground will re-freeze. This cycle might be repeated for several years in a row if summer temperatures continue to remain cool.

19. As a result of increased volcanic activity, a large amount of volcanic ash and gases would be in the air. These particles would have reflected many of the sun's rays back into space and kept them from reaching the earth, producing a cooling effect on the earth.

20. During the Flood, there had been huge volcanic eruptions in the ocean and the release of steam and heat from deep breaks in the oceanic crust. This would have caused the ocean water to become warmer. Warm ocean water would have heated the air above and evaporated much more water, resulting in more clouds, rain, and snow.

Water & Weather Test Answer Key Level 1 & 2

Test 1 Level 1

1. sedimentary
2. sea
3. decay
4. oldest, bottom
5. graveyards
6. evolutionary
7. rivers, estuaries, oceans
8. temperature
9. shelf
10. plains
11. slope
12. convection
13. wind
14. climate
15. runoff
16. evaporation, condensation
17. rain
18. layers
19. cumulonimbus
20. tilt
21. heating, weight
22. prevailing winds
23. rain
24. air pressure
25. lightweight
26. front
27. thunderstorms, tornadoes
28. rotation
29. tropical
30. True
31. True
32. False
33. False
34. True
35. False
36. True
37. True
38. True
39. False
40. When an organism changes into a fossil, the minerals in the dead organism are replaced by minerals in the surrounding mud or water and cemented together.
41. Sediments were laid down by water in flat layers that eventually harden into rock. The shape of the layers can also be altered by water. An example of this would be heavy rains eroding the layers to form canyons.
42. Water molecules evaporate from the earth after being warmed by the sun. As the water vapor cools at higher altitudes, the vapor condenses into liquid water, which may fall to the earth in the form of rain or snow.
43. By dividing the weight (or mass) of the object by its volume. (density = mass/volume)
44. Pure science provides explanations for things that have been observed in nature, while technology is about making things that have a useful purpose.

Test 1 Level 2

1. b
2. c
3. c
4. c
5. b
6. b
7. a
8. c
9. b
10. c
11. c
12. a
13. c
14. a
15. a
16. b
17. d
18. c
19. b
20. a
21. c
22. b
23. c
24. d
25. a
26. a
27. d
28. b
29. b
30. b
31. Rapid burial by layers of sedimentary deposits would have kept remains of plants and animals from being eaten, and the sediment would have quickly cut off the oxygen supply. Minerals necessary for fossilization would have been dissolved in the water. The weight of the water and deep layers of sediments would have resulted in pressure on the remains.

32. The progression of the evolution of all living things, along with the ages in which different groups evolved.

33. The progression of events that occurred during the Flood

34. Temperature, amount of sunlight, amount of water pressure, food supply

35. Sunlight can only penetrate about 650 feet.

36. Water molecules evaporate from the earth after being warmed by the sun. As the water vapor cools at higher altitudes, the vapor condenses into liquid water which may fall to the earth in the form of rain or snow.

37. The warm humid air tends to rise in the air. The higher it rises, the cooler the air becomes. Eventually the moisture in the air condenses into tiny droplets and forms a cloud.

38. As the air becomes colder, it cannot hold much water vapor. Since very cold air is also dry, it will not produce much snow. Increased snowfall occurs when the air is not too cold. Snow and ice accumulate when summer temperatures are cool enough that much of the snow doesn't melt. Then the next year's snows will add another layer of ice to what is already on the ground. Also, any liquid water on the ground will re-freeze. This cycle might be repeated for several years in a row if summer temperatures continue to remain cool.

39. For the answers to the Bonus Question, see diagram on page 82.

The Universe Quiz Answer Keys Level 1 & 2

Quiz One, Level 1 Chapters 1-3

1. asteroid belt
2. TNO
3. solar system
4. galaxy
5. natural world
6. nebula theory
7. telescopes
8. parallax effect
9. supernatural
10. vacuum
11. Mercury, Venus, Earth, Mars
12. Jupiter, Saturn, Uranus, and Neptune
13. AU stands for Astronomical Unit
14. The distance light can travel through space in a year
15. red
16. They concluded that the sun was the center of the solar system, and that the earth and planets revolved around the sun.
17. light, space, mass, and gravity
18. special theory of relativity and general theory of relativity
19. False
20. True
21. True
22. True
23. False
24. His first theory showed that Newton's three laws of motion don't always work when objects approach the speed of light. His second theory showed that Newton's law of gravitation doesn't always work when the gravitational force becomes very strong.

Quiz Two, Level 1 Chapters 4-7

1. predictable, elliptical
2. elongated
3. gravitational
4. friction
5. predictable
6. masses, distance
7. inertia
8. convex, magnifying
9. refraction
10. mirrors
11. Galileo
12. Copernicus believed the planets followed a circular orbit around the sun.
13. No
14. The moon doesn't have an atmosphere of air around it, so there would be no air resistance to push up on the feather.
15. scientific tests
16. False
17. True
18. True
19. False
20. True
21. They both contain two convex lenses in an adjustable tube. One of the lenses is a thick lens and the other one is a thin lens. The lenses are reversed in a microscope.

Quiz Three, Level 1 Chapters 8-11

1. Archimedes
2. Newton, reaction
3. air
4. meteorites
5. photosynthesis, carbon dioxide
6. oxygen
7. oxygen, carbon dioxide
8. tides
9. eclipse
10. oxygen
11. 1903
12. People were too heavy and their muscles were too weak to fly.
13. National Aeronautics and Space Administration
14. Venus is not tilted on its axis.
15. The earth's atmosphere traps and holds heat from the sun.
16. False
17. True
18. True
19. False
20. True
21. The earth is not too close or too far away from the sun to upset

the water cycle. There is the right amount of gravitational force to hold the earth's atmosphere in place. There is enough oxygen to live, but not so much that fires would be uncontrollable. Air pressure is enough to balance internal pressures. The earth's magnetic field acts as a shield for many kinds of dangerous particles and rays that bombard the earth from outer space.

Quiz Four, Level 1 Chapters 12-15

1. axis
2. south pole
3. Venus, Mars, Jupiter
4. Mercury, Earth
5. Jovian
6. hydrogen, helium
7. rings
8. larger
9. frozen
10. Because there is no air on the moon
11. We can only see the part of the moon that is illuminated by the sun.
12. It means that both the moon and the sun appear to be the same size in the sky.
13. Venus
14. People began to think that the best way to find out what is true was not to look at what God revealed in His Word, but to use human reason. There was a shift from "revelation to reason."
15. True
16. False
17. False
18. True
19. True
20. Some scientists have been considering evidence that there were once advanced civilizations. The pre-Flood people were probably very intelligent and had developed many technologies. Some of this information would have been

passed on to Noah and his family who survived the Flood.

Quiz Five, Level 1 Chapters 16-20

1. spectrum
2. electromagnetic
3. ultraviolet waves, infrared waves
4. sundial
5. magnetic field
6. stable, variable
7. spectroscope
8. constellations
9. supernova, nebula
10. Electromagnetic waves travel at the speed of light.
11. Very short electromagnetic waves are more dangerous.
12. Greenwich, England
13. 88 constellations
14. Orion, the Pleiades, and the Bear or Ursa Major
15. color, temperature, size, chemical composition, spectra, and the kinds of radiations they emit
16. True
17. False
18. True
19. False
20. True
21. One explanation proposes that the entire universe began with the big bang and then all the stars, planets, and moons gradually formed over billions of years. The other explanation proposes that God supernaturally created the earth, the rest of the solar system, and the stars according to His plan and design.

Quiz One, Level 2 Chapters 1-3

1. c
2. b
3. b
4. a
5. d
6. e
7. a

8. b
9. a
10. c
11. The solar system is made up of the sun and the planets that orbit the sun, along with everything else held in place by the sun's gravity.
12. A galaxy is made up of millions or billions of stars that are bound together by gravitational attraction.
13. The universe is a term used to include everything that exists in the natural world.
14. The Milky Way is the name given to the galaxy that contains our solar system.
15. The Local Group is the name of a cluster of a few dozen galaxies that contains our Milky Way Galaxy.
16. The speed of light in a vacuum
17. His first theory showed that Newton's three laws of motion don't always work when objects approach the speed of light. His second theory showed that Newton's law of gravitation doesn't always work when the gravitational force becomes very strong.
18. They are dependable throughout the universe as long as they are viewed from a location that is assumed to be flat and not moving.
19. A red shift in the light spectra of galaxies
20. He collected spectra of light from 46 galaxies and noted that there was always a red shift in the colors of the visible spectrum that came from these galaxies.

Quiz Two, Level 2 Chapters 4-7

1. a
2. a
3. d
4. b

5. c
6. d
7. d
8. d
9. b
10. a
11. The tilt of the earth on its axis
12. Their masses and the distance between them
13. There is a greater pull from gravity on the heavy rock, but at the same time, the heavy rock's inertia has more resistance to moving than the lightweight rock. These two components cancel each other.
14. They both contain two convex lenses in an adjustable tube. One of the lenses is a thick lens and the other one is a thin lens. The lenses are reversed in a microscope.
15. No

Quiz Three, Level 2 Chapters 8-11

1. d
2. a
3. c
4. d
5. c
6. a
7. a
8. b
9. a
10. d
11. b
12. d
13. Hindenburg
14. For every action, there is an equal and opposite reaction.
15. Animals breathe out carbon dioxide and water, which plants use to make food. Plants give off oxygen which animals breathe in.
16. Rockets have been used to carry spacecraft and satellites into space. Rockets have enabled scientists to explore the solar system, send communication and weather satellites into orbit, and put a series of GPS satellites into place.
17. Troposphere, stratosphere, mesosphere, thermosphere, and exosphere
18. When there is no air pressure, the water boils away, even in cold temperatures.
19. Although asteroids and rocks from space are often attracted to the earth by gravity, they usually burn up as they begin to hit the oxygen in the air.
20. It is not tilted on its axis.

Quiz Four, Level 2 Chapters 12-15

1. d
2. a
3. b
4. c
5. d
6. d
7. b
8. c
9. b
10. d
11. Because the moon rotates on its axis once while it makes one orbit around the earth
12. From the earth, we can only see the part of the moon that is illuminated by the sun.
13. The earth's gravitational pull holds the moon in its orbit around the earth.
14. It means that both the moon and the sun appear to be the same size in the sky.
15. People began to think that the best way to find out what is true was not to look at what God revealed in His Word, but to use human reason. There was a shift from "revelation to reason."

Quiz Five, Level 2 Chapters 16-20

1. d
2. d
3. a
4. d
5. b
6. c
7. a
8. c
9. d
10. b
11. Sound waves cannot travel through space. They also travel much slower than electromagnetic waves and differ in how they vibrate.
12. One explanation proposes that the entire universe began with the big bang and then all the stars, planets, and moons gradually formed over billions of years. The other explanation proposes that God supernaturally created the earth, the rest of the solar system, and the stars according to His plan and design.
13. Infrared waves
14. They travel through space at the speed of light.
15. Scientists believe that energy is released by nuclear fusion as hydrogen nuclei are converted into helium nuclei.

The Universe Test Answer Key

Test 1 Level 1

1. TNO
2. solar system
3. galaxy
4. natural world
5. nebula theory
6. supernatural
7. predictable, elliptical
8. gravitational
9. friction
10. Galileo
11. Newton, reaction
12. meteorites
13. oxygen, carbon dioxide
14. eclipse

15. oxygen
16. Jovian
17. hydrogen, helium
18. rings
19. larger
20. frozen
21. spectrum
22. electromagnetic
23. magnetic field
24. False
25. True
26. True
27. False
28. True
29. True
30. True
31. True
32. False
33. False
34. AU stands for Astronomical Unit
35. A light year is the distance light can travel through space in a year.
36. Special theory of relativity and general theory of relativity
37. His first theory showed that Newton's three laws of motion don't always work when objects approach the speed of light. His second theory showed that Newton's law of gravitation doesn't always work when the gravitational force becomes very strong.
38. No
39. The moon doesn't have an atmosphere of air around it, so there would be no air resistance to push up on the feather.
40. People were too heavy and their muscles were too weak to fly.
41. The earth's atmosphere traps and holds heat from the sun.
42. The earth is not too close or too far away from the sun to upset the water cycle. There is the right amount of gravitational force to hold the earth's atmosphere in place. There is enough oxygen to live, but not so much that fires would be uncontrollable. Air pressure is enough to balance internal pressures. The earth's magnetic field acts as a shield for many kinds of dangerous particles and rays that bombard the earth from outer space.

43. It means that both the moon and the sun appear to be the same size in the sky.

44. People began to think that the best way to find out what is true was not to look at what God revealed in His Word, but to use human reason. There was a shift from "revelation to reason."

45. Orion, the Pleiades, and the Bear or Ursa Major

46. Color, temperature, size, chemical composition, spectra, and the kinds of radiations they emit

Test 1 Level 2

1. b
2. a
3. d
4. a
5. b
6. c
7. a
8. d
9. c
10. d
11. d
12. b
13. c
14. c
15. a
16. a
17. a
18. d
19. b
20. c
21. d
22. b
23. c
24. d
25. d
26. a
27. d
28. b
29. c
30. b
31. The speed of light in a vacuum
32. His first theory showed that Newton's three laws of motion don't always work when objects approach the speed of light. His second theory showed that Newton's law of gravitation doesn't always work when the gravitational force becomes very strong.
33. He collected spectra of light from 46 galaxies and noted that there was always a red shift in the colors of the visible spectrum that came from these galaxies.
34. their masses and the distance between them
35. There is a greater pull from gravity on the heavy rock, but at the same time, the heavy rock's inertia has more resistance to moving than the lightweight rock. These two components cancel each other.
36. When there is no air pressure, the water boils away, even in cold temperatures.
37. People began to think that the best way to find out what is true was not to look at what God revealed in His Word, but to use human reason. There was a shift from "revelation to reason."
38. Scientists believe that energy is released by nuclear fusion as hydrogen nuclei are converted into helium nuclei.

Master Supply List

for Use with

Science Starters: *Elementary General Science & Astronomy*

Water & Weather
Semester Supply List
Common Household Items

- 1-cup measure
- 2-liter bottles (empty)
- Air thermometer
- Baking soda
- Ball
- Balloons
- Bowl
- Bowls (or foam trays)
- Cardboard boxes
- Cardboard roll from toilet paper or wrapping paper
- Chicken wing bones
- Clay or play dough (not too sticky)
- Clear bottle with narrow neck
- Clear tape
- Colored makers
- Colored marker (dark)
- Colored water frozen into cubes
- Compass
- Container for mixing plaster (disposable)
- Container of cool water
- Container of hot and ice water
- Container to catch dripping water
- Containers (in which bones can soak)
- Containers for mixing
- Cool water
- Cup
- Cylindrical object to use in rolling over clay
- Diagram of rock strata, intrusive rocks, and canyon in the appendix
- Dinosaur books
- Dirt
- Dirt containing little sticks (small amout)
- Distilled water
- Drinking glass
- Drinking straw
- Duct or masking tape
- File folder or other stiff paper
- Flash light
- Food coloring
- Freezer
- Glasses or plastic cups
- Global map

- Hair dryer or fan
- Hammer
- Hot water
- Ice cube
- Internet or other references about the Mississippi River, its tributaries, and the Louis and Clark expedition
- Lamps (electric)
- Large pan
- Large plastic pan
- Leaf (not too fragile)
- Lettuce leaves
- Lids of shoebox
- Liquid detergent or water container with a release spout (empty)
- Map of the United States showing major rivers and lakes (see appendix)
- Marker
- Measuring cup (milliliters)
- Measuring tape
- Meat tray (small)
- Metal cans (shiney)
- Metric ruler
- Milk carton
- Nail
- National weather maps from a newspaper or the Internet collected over a 5-day period
- Paint brush
- Pans (shallow)
- Paper
- Paper clip
- Paper plates
- Paper towels
- Pattern for arrow (appendix)
- Pen or pencil
- Pencil with eraser
- Pitcher of warm water
- Plaster of Paris
- Plastic glasses
- Plastic wrap
- Potato wedges
- Recipe for play dough
- References about weather
- Rocks
- Ruler (Metric/English)
- Ruler (small)
- Safety goggles

- ❏ Salt
- ❏ Salt solutions (concentrated, colored)
- ❏ Scissors
- ❏ Shaving cream (not gel)
- ❏ Sheets of paper
- ❏ Soda bottle (empty)
- ❏ Sponges
- ❏ Spoon
- ❏ Stir sticks
- ❏ Stirer
- ❏ Stirring spoons
- ❏ Straight pin (or hat pin)
- ❏ Straws
- ❏ String
- ❏ Strips of paper
- ❏ Styrofoam ball
- ❏ Sugar
- ❏ Tape
- ❏ Teaspoon
- ❏ Test tubes
- ❏ Thermometer
- ❏ Tin can (empty)
- ❏ Toothpick
- ❏ Toy (small, plastic)
- ❏ Up-to-date references about submersibles and the deep ocean zones.
- ❏ Vasoline or oil
- ❏ Vinegar
- ❏ Washed sand with small amount of washed gravel
- ❏ Water
- ❏ White paper
- ❏ Wire screens a little larger than the trays
- ❏ Wooden block
- ❏ Ziplock baggie

List courtesy of: **InvestigateThePossibilities.org**

Visit the site for more information and specialty items.

Water & Weather Supply List by Investigation

Investigation #1: In the Beginning...God Created Dinosaurs
Gather These Things:

- ❏ 1½ - 2 cups of clay or play dough (not too sticky)
- ❏ Paper plates
- ❏ Paper towels
- ❏ Disposable container for mixing plaster
- ❏ Container of cool water
- ❏ 1-cup measure
- ❏ Stirring spoons
- ❏ Plaster of Paris (plain)
- ❏ Vasoline or oil
- ❏ Small meat tray
- ❏ Plastic wrap
- ❏ Small plastic toy or other object to represent a foot (be creative; each "footprint" should be at least 0.5 cm wide)
- ❏ Wooden block
- ❏ Leaf (not too fragile)
- ❏ Cylindrical object to use in rolling over clay
- ❏ Lids of shoeboxes
- ❏ Metric ruler
- ❏ Teacher's book recipe for play dough

Investigation #2: Making a Big Impression!
Gather These Things:

- ❏ (3) bowls (or foam trays)
- ❏ (3) wire screens a little larger than the trays
- ❏ Containers for mixing
- ❏ 1½ Cups of cool water
- ❏ 3 cups of hot water
- ❏ ¼ cup salt
- ❏ ¼ cup sugar
- ❏ Plaster of Paris
- ❏ Stir sticks
- ❏ Baking soda
- ❏ Vinegar
- ❏ Spoon
- ❏ 2 Chicken wing bones
- ❏ (3) sponges
- ❏ 2 containers (in which bones can soak)

Investigation #3: No Bones About It!
Gather These Things:

- ❏ Old toothbrush
- ❏ Screwdrivers
- ❏ Popsicle stick
- ❏ Paper towels
- ❏ Paper plate
- ❏ Toothpicks
- ❏ Prepared rock biscuit with "fossil" inside
- ❏ Chocolate chip cookies

Investigation #4: Digging In and Reconstructing Fossils
Gather These Things:

- ❏ Measuring tape
- ❏ Several 1 meter strips of paper to lay on floor
- ❏ Reference materials about dinosaurs
- ❏ Ziplock baggies one for each student
- ❏ Rocks – several for each student
- ❏ Lettuce leaves
- ❏ Vinegar

Investigation #5: Can Rocks tell Time?
Gather These Things:

- ❏ Empty liquid detergent or water container with a release spout
- ❏ Toothpick
- ❏ Food coloring
- ❏ Container to catch dripping water

Investigation #6: Leave No Stone Unturned
Gather These Things:

- ❏ Diagram of rock strata, intrusive rocks, and canyon in the appendix

Investigation #7: Just How Salty is the Ocean?
Gather These Things:

- ❏ Three glasses or plastic cups (at least 12 oz size)
- ❏ Distilled water (at least 24 oz)
- ❏ Three wedges of a potato, about same size
- ❏ Straws
- ❏ Salt
- ❏ Cool water
- ❏ Test tubes
- ❏ Sheets of paper
- ❏ Teaspoon
- ❏ 3 different bottles of food coloring

Investigation #8: Ocean Zones—from Light to Dark Places
Gather These Things:

- ❏ Milk carton with 3 holes down one side in a vertical line
- ❏ Duct or masking tape
- ❏ Water
- ❏ Large plastic pan

Investigation #9: From Shelf to Shelf and In-Between
Gather These Things:

- ❏ Up-to-date references about submersibles and the deep ocean zones.

Investigation #10: Currents in the Ocean
Gather These Things:

- ❏ Global map
- ❏ Colored water frozen into cubes
- ❏ Large pan
- ❏ Pitcher of warm water

Investigation #11: Where the Rivers Run into the Sea
Gather These Things:

- ❏ Map of the United States showing major rivers and lakes (see appendix)
- ❏ Colored makers
- ❏ Internet or other references about the Mississippi River, its tributaries, and the Louis and Clark expedition

Investigation #12: Evaporation, Condensation, and the Water Cycle
Gather These Things:

- ❏ Concentrated, colored salt solutions
- ❏ Paint brush
- ❏ Paper
- ❏ Colored markers
- ❏ Drinking glass
- ❏ Ice cubes and water
- ❏ Bowl
- ❏ Plastic wrap

Investigation #13: God's System of Purifying Water
Gather These Things:

- ❏ Small amount of dirt containing little sticks
- ❏ Water

- ❏ Cup to mix dirt and water
- ❏ Cup to collect water
- ❏ Stirer
- ❏ Hammer
- ❏ Nail
- ❏ Empty tin can
- ❏ Washed sand with small amount of washed gravel

Investigation #14: Weather or Not!
Gather These Things:

- ❏ Shaving cream (not gel)
- ❏ 3 paper plates
- ❏ Pen or pencil
- ❏ Ice cube
- ❏ Safety goggles
- ❏ Shiny metal cans
- ❏ Clear bottle with narrow neck
- ❏ References about weather

Investigation #15: Day and Night, Summer and Winter
Gather These Things:

- ❏ Clear tape
- ❏ String
- ❏ Ball
- ❏ Colored tape
- ❏ Flashlight
- ❏ Paper clip
- ❏ Styrofoam ball
- ❏ Marker
- ❏ White paper

Investigation #16: The Temperature's Rising
Gather These Things:

- ❏ (2) 8-oz plastic glasses
- ❏ Water
- ❏ Dirt
- ❏ 2 electric lamps (or other light that provides heat)
- ❏ 2 Thermometers (same kind)
- ❏ Round-shaped balloons
- ❏ Empty soda bottle
- ❏ Container of hot and ice water
- ❏ Dark colored marker
- ❏ Flash light
- ❏ Empty cardboard roll from toilet paper or

wrapping paper
- ❏ Small ruler

Investigation #17: Weather Instruments
Gather These Things:

- ❏ File folder or other stiff paper
- ❏ Scissors
- ❏ Drinking straw
- ❏ Straight pin (or hat pin)
- ❏ Pencil with eraser
- ❏ Tape
- ❏ Hair dryer or fan
- ❏ Pattern for arrow (appendix)
- ❏ Compass
- ❏ 1-liter plastic soda bottle
- ❏ Marker
- ❏ Food coloring
- ❏ Metric/English ruler
- ❏ Air thermometer

Investigation #18: Forecasting the Weather
Gather These Things:

- ❏ National weather maps from a newspaper or the Internet collected over a 5-day period

Investigation #19: From Gentle Breezes to Dangerous Winds
Gather These Things:

- ❏ 2 empty 2-liter bottles
- ❏ Tape

Investigation #20: Climate Change
Gather These Things:

- ❏ 2 shallow pans (same size and shape)
- ❏ 2 cardboard boxes (same size and shape and larger than the pans)
- ❏ 1 small lamp (remove shade)
- ❏ Scissors
- ❏ Water
- ❏ Freezer
- ❏ Thermometer
- ❏ Measuring cup (milliliters)
- ❏ Metric ruler

List courtesy of: **InvestigateThePossibilities.org**
Visit the site for more information and specialty items.

The Universe
Semester Supply List
● **Common Household Items**

- ❑ 35 mm film canister
- ❑ Ball
- ❑ Balloon
- ❑ Bucket (small with handle)
- ❑ Cardboard
- ❑ Constellation diagrams
- ❑ Cookie sheet (flat)
- ❑ Cup (plastic, small)
- ❑ Dime
- ❑ Drinking cups (paper, cone-shaped)
- ❑ Effervescent antacid tablet
- ❑ Flashlight
- ❑ Flexible pipe insulation tubing
- ❑ Flexible tubing
- ❑ Flour
- ❑ Globe
- ❑ Glue
- ❑ Index card (4" x 6")
- ❑ Magnifying (convex) lenses
- ❑ Marbles
- ❑ Marker (colored)
- ❑ Markers (silver)
- ❑ Marshmallows (large)
- ❑ Meter stick or long flat board
- ❑ Other objects to observe falling
- ❑ Paper
- ❑ Paper (4" x 4")
- ❑ Paper (heavy, black)
- ❑ Paper (lined)
- ❑ Paper (strips)
- ❑ Paper clips (large)
- ❑ Paprika (red)
- ❑ Pattern for paper airplane
- ❑ Pen
- ❑ Pencil
- ❑ Pennies (40)
- ❑ Pictures of space (Internet, books, and magazines)
- ❑ Prism
- ❑ Quarter
- ❑ References (Internet or library)
- ❑ Rocky planets (resource sources)
- ❑ Rope
- ❑ Ruler or straight edge

- ❑ Scissors
- ❑ Scrabble letter tiles
- ❑ Shoebox
- ❑ Source of light (strong)
- ❑ spray paint (flat black)
- ❑ Sticky note
- ❑ Straight pole or stick
- ❑ String
- ❑ Sunscreen (different brands)
- ❑ Supplies to make a model of an eclipse
- ❑ Tape (clear)
- ❑ Tape measure (metric)
- ❑ Trash bag
- ❑ Typing paper (8.5 x 11)
- ❑ Watch
- ❑ Water
- ❑ White surface

List courtesy of: **InvestigateThePossibilities.org**

Visit the site for more information and specialty items.

The Universe Supply List by Investigation

Investigation #1: What Is the Universe?
Gather These Things:
- ❏ 3 sheets of 8½ X 11 typing paper
- ❏ Clear tape
- ❏ 40 pennies
- ❏ Pen or pencil
- ❏ Ruler or straight edge

Investigation #2: Spreading Out the Heavens
Gather These Things:
- ❏ Large round rubber balloon
- ❏ Colored marker
- ❏ Bag of large marshmallows

Investigation #3: The Strange Behavior of Space and Light
Gather These Things:
- ❏ Large round ball
- ❏ Strips of paper
- ❏ Tape
- ❏ Piece of string
- ❏ Metric ruler
- ❏ Colored marker

Investigation #4: Kepler's Clockwise Universe
Gather These Things:
- ❏ A large round ball
- ❏ Strips of paper
- ❏ Tape
- ❏ Piece of string
- ❏ Metric ruler
- ❏ Colored marker

Investigation #5: Invisible Forces in the Universe
Gather These Things:
- ❏ Flat sheet of paper
- ❏ Tape
- ❏ Flexible tubing (cut in half lengthwise or cardboard cut in half lengthwise)
- ❏ Marble
- ❏ Meter stick or long flat board
- ❏ Other objects to observe falling, examples: small pebble, an other rock, foam ball.

Investigation #6: Galileo and Inertia
Gather These Things:
- ❏ Small bucket with a handle
- ❏ Short rope
- ❏ Water
- ❏ Flexible pipe insulation tubing, cut in half lengthwise
- ❏ Marbles
- ❏ Tape

Investigation #7: Making Telescopes
Gather These Things:
- ❏ 2 magnifying (convex) lenses
- ❏ Lined paper

Investigation #8: History of Flight
Gather These Things:
- ❏ 8½ x 11 sheet of paper
- ❏ Large paper clips
- ❏ Pattern for paper airplane (see appendix of Student Journal)
- ❏ Metric tape measure

Investigation #9: Rockets and Space Exploration
Gather These Things:
- ❏ 4" x 4" piece of paper
- ❏ 2 cone-shaped paper drinking cups (use one for cutting out the fins)
- ❏ Clear tape
- ❏ Scissors
- ❏ Effervescent antacid tablet
- ❏ 35 mm film canister with lid that fits inside canister
- ❏ Marker
- ❏ Trash bag

Investigation #10: The Earth in Space
Gather These Things:
- ❏ Globe of the earth, mounted from its axis
- ❏ A ball or balloon (about ¼ the size of the globe)
- ❏ Flashlight
- ❏ Sticky note
- ❏ Axis-mounted earth globe (may be viewed at a library if one is not available elsewhere)

Investigation #11: The Earth's Atmosphere
Gather These Things:

- Small plastic cup
- 4" x 6" index card
- Water

Investigation #12: The Moon
Gather These Things:

- Flat cookie sheet
- A Lincoln penny
- A quarter
- Paper
- Pencil
- Flour
- A metric tape
- Red paprika
- Marble
- Dime
- Clear tape

Investigation #13: The Rocky Planets: Mercury, Venus, Earth, and Mars
Gather These Things:

- Up-to-date reference sources and Internet sources of information about the rocky planets.

Investigation #14: Planets: Mars and Martians
Gather These Things:

- Internet or library materials to reference.

Investigation #15: The Jovian Planets: Jupiter, Saturn, Uranus, and Neptune
Gather These Things:

- Internet or library materials to reference.

Investigation #16: The Sun and Its Light
Gather These Things:

- A few different bottles of sunscreen
- A prism
- White surface
- Source of light (strong)

Investigation #17: The Sun and the Earth Relationship
Gather These Things:

- Straight pole or stick
- Metric measuring tape
- Watch
- Supplies to make a model of an eclipse
- Ball / flashlight

Investigation #18: The Constellations
Gather These Things:

- Supplies for shoebox constellations: Large shoebox, heavy black paper, glue, tape, cardboard, silver markers, constellation diagrams (from appendix and other sources)
- Flat black spray paint

Investigation #19: A Great Variety in Space
Gather These Things:

- Pictures from the Internet or from books and magazines of space.

Investigation #20: Chaos or a Creator?
Gather These Things:

- Letter tiles from a game of Scrabble that spell the word "CREATION"

List courtesy of: **InvestigateThePossibilities.org**
Visit the site for more information and specialty items.

Biblical Beginnings

Ages 3, 4, & 5 years old

Package Includes: *A is for Adam; A is for Adam DVD; D is for Dinosaur; D is for Dinosaur DVD; N is for Noah; Noah's Ark Preschool; God Made the World & Me; Dinosaurs Stars of the Show; Big Thoughts for Little Thinkers: The Gospel; The Mission; The Scripture; The Trinity; Creation Story for Children; When Dragons Hearts Were Good; Dinosaur by Design; Biblical Beginnings Preschool , Parent Lesson Planner*

14 Book, 2 DVD Package	978-0-89051-887-8	**$187.84**
PLP Only (168 Pages)	978-0-89051-885-4	**$14.99**

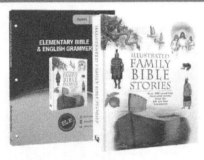

Elementary Bible & English Grammar

1 year: 3rd–6th grade

Package Includes: *Illustrated Family Bible Stories and Parent Lesson Planner*

2 Book Package	978-0-89051-852-6	**$39.99**
PLP Only (130 Pages)	978-0-89051-851-9	**$14.99**

Elementary Geography & Cultures

1 year: 3rd–6th grade

Package Includes: *Children's Atlas of God's World, Passport to the World, & Parent Lesson Planner*

3 Book	978-0-89051-814-4	**$49.99**
PLP Only (130 Pages)	978-0-89051-808-3	**$14.99**

Elementary World History

1 year: 5th–8th grade

Package Includes: *The Big Book of History; Noah's Ark: Thinking Outside the Box (book and DVD); Parent Lesson Planner*

3 Book, 1 DVD Package	978-0-89051-815-1	**$64.99**
PLP Only (260 Pages)	978-0-89051-809-0	**$15.99**

Elementary Zoology

1 year: 3th–6th grade

5 Book Package Includes: *World of Animals; Dinosaur Activity Book; The Complete Aquarium Adventure; The Complete Zoo Adventure; Parent Lesson Planner*

5 Book Package	978-0-89051-747-5	**$85.99**
PLP Only (178 Pages)	978-0-89051-724-6	**$14.99**

Science Starters: Elementary Chemistry & Physics

1 year: 3rd–6th grade

7 Book Package Includes: *Matter- Student, Student Journal, and Teacher; Energy- Student, Teacher, and Student Journal; Parent Lesson Planner*

7 Book Package   978-0-89051-749-9   **$54.99**
PLP Only (100 Pages)   978-0-89051-726-0   **$8.99**

Science Starters: Elementary General Science & Astronomy

1 year: 3rd–6th grade

7 Book Package Includes: *Water & Weather- Student Journal and Teacher; The Universe- Student, Teacher, & Student Journal; Parent Lesson Planner*

7 Book Package   978-0-89051-816-8   **$54.99**
PLP Only (98 Pages)   978-0-89051-810-6   **$8.99**

Science Starters: Elementary Physical & Earth Science

1 year: 3rd–6th grade

6 Book Package Includes: *Forces & Motion- Student, Student Journal, and Teacher; The Earth- Student, Teacher & Student Journal; Parent Lesson Planner*

6 Book Package   978-0-89051-748-2   **$51.99**
PLP Only (94 Pages)   978-0-89051-725-3   **$8.99**

Applied Engineering: Studies of God's Design In Nature

1 year Engineering (1 credit): 7th–9th grade

Package Includes: *Made in Heaven, Champions of Invention, Discovery of Design, Parent Lesson Planner*

4 Book Package   978-0-89051-812-0   **$53.99**
PLP Only (260 Pages)   978-0-89051-806-9   **$17.99**

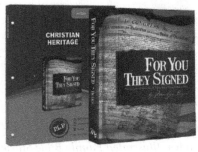

Christian Heritage

1 year Christian Heritage (1 credit): 9th–12th grade

Package Includes: *For You They Signed; Parent Lesson Planner*

2 Book Package   978-0-89051-769-7   **$50.99**
PLP Only (240 Pages)   978-0-89051-746-8   **$15.99**

Christian History: Biographies of Faith

1 year History (1 credit): 7th–9th grade

6 Book Package Includes: *Life of John Newton, Life of Washington, Life of Andrew Jackson, Life of John Knox, Life of Luther, Parent Lesson Planner*

6 Book Package   978-0-89051-847-2   **$101.99**
PLP Only (176 Pages)   978-0-89051-850-2   **$16.99**

Advanced Pre-Med Studies

1 year Biology (½ credit): 10th–12th grade

Package Includes: *Building Blocks in Life Science; The Genesis of Germs; Body by Design; Exploring the History of Medicine; Parent Lesson Planner*

5 Book Package 978-0-89051-767-3		**$79.99**
PLP Only (238 pages) 978-0-89051-744-4		**$17.99**

Basic Pre-Med

1 year Biology (½ credit): 8th–9th grade

Package Includes: *The Genesis of Germs; The Building Blocks in Life Science; Parent Lesson Planner*

3 Book Package 978-0-89051-759-8		**$45.99**
PLP Only (120 Pages) 978-0-89051-736-9		**$12.99**

Cultural Issues: Creation/Evolution and the Bible

1 year Apologetics (½ credit): 10th–12th grade

Package Includes: *New Answers Book 1; New Answers Book 2; Parent Lesson Planner*

3 Book Package 978-0-89051-846-5		**$44.99**
PLP Only (208 Pages) 978-0-89051-849-6		**$14.99**

Apologetics in Action

1 year Apologetics (1 credit): 10th–12th grade

Package Includes: *How Do I Know the Bible is True volumes 1 & 2; Demolishing Supposed Bible Contradictions Volumes 1 & 2; Parent Lesson Planner*

5 Book Package 978-0-89051-848-9		**$70.99**
PLP Only (176 Pages) 978-0-89051-839-7		**$14.99**

Biblical Archaeology

1 year Bible/Archaeology (1 credit): 10th–12th grade

Package Includes: *Unwrapping the Pharaohs; Unveiling the Kings of Israel; The Archaeology Book; Parent Lesson Planner*

4 Book Package 978-0-89051-768-0		**$99.99**
PLP Only (244 Pages) 978-0-89051-745-1		**$17.99**

Intro to Biblical Greek

1 year Foreign Language (½ credit): 10th–12th grade

Package Includes: *It's Not Greek to Me DVD & Parent Lesson Planner*

1 Book, 1 DVD Package 978-0-89051-818-2		**$33.99**
PLP Only (132 Pages) 978-0-89051-817-5		**$13.99**

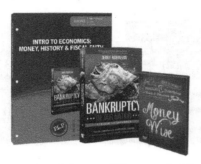

Intro to Economics: Money, History, & Fiscal Faith

1/2 year Economics (½ credit): 10th–12th grade

Package Includes: *Bankruptcy of Our Nation, Money Wise DVD, Parent Lesson Planner*

2 Book, 4 DVD Package	978-0-89051-811-3	**$57.99**
PLP Only (230 Pages)	978-0-89051-805-2	**$13.99**

Life Science: Origins & Scientific Theory

1 year Palentology (1 credit): 7th–9th grade

Package Includes: *Evolution: the Grand Experiment, Teacher Guide, DVD; Living Fossils, Teacher Guide, DVD; Parent Lesson Planner*

5 Book, 2 DVD Package	978-0-89051-761-1	**$144.99**
PLP Only (60 Pages)	978-0-89051-738-3	**$5.99**

Natural Science: The Story of Origins

1 year Natural Science (½ credit): 10th–12th grade

Package Includes: *Evolution: The Grand Experiment; Evolution: The Grand Experiment Teacher's Guide, Evolution: The Grand Experiment DVD; Parent Lesson Planner*

3 Book, 1 DVD Package	978-0-89051-762-8	**$71.99**
PLP Only (35 Pages)	978-0-89051-739-0	**$4.99**

Paleontology: Living Fossils

1 year Paleontology (½ credit): 10th–12th grade

Package Includes: *Living Fossils, Living Fossils Teacher Guide, Living Fossils DVD; Parent Lesson Planner*

3 Book, 1 DVD Package	978-0-89051-763-5	**$66.99**
PLP Only (31 Pages)	978-0-89051-740-6	**$4.99**

Survey of Astronomy

1 year Astronomy (1 credit): 10th–12th grade

Package Includes: *The Stargazers Guide to the Night Sky; Our Created Moon; Taking Back Astronomy; Our Created Moon DVD; Created Cosmos DVD; Parent Lesson Planner*

4 Book, 2 DVD Package	978-0-89051-766-6	**$113.99**
PLP Only (250 Pages)	978-0-89051-743-7	**$17.99**

Survey of Science History & Concepts

1 year Science (1 credit): 10th–12th grade

Package Includes: *The World of Mathematics; The World of Physics; The World of Biology; The World of Chemistry; Parent Lesson Planner*

5 Book Package	978-0-89051-764-2	**$72.99**
PLP Only (216 Pages)	978-0-89051-741-3	**$16.99**

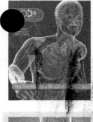

Breathtaking Respiratory System, The
Dr. Lainna Callentine

An elementary-level exploration of the human body's respiratory system, focused on structures, function, diseases, and God's design. Created by pediatrician and homeschool mom, Dr. Lainna Callentine.
C 978-0-89051-862-5 **$15.99** U.S.

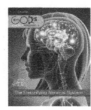

Electrifying Nervous System, The
Dr. Lainna Callentine

Learn interesting and important facts about why you sleep, what foods can superpower your brain functions, and much more in a wonderful exploration of the brain and how it controls your body!
C 978-0-89051-833-5 **$15.99** U.S.

Complex Circulatory System, The
Dr. Lainna Callentine

Focuses on the heart, blood, and blood vessels that make up the body's circulatory system. Beyond the basics of how and why the body works, students learn God's amazing and deliberate design.
C 978-0-89051-908-0 **$15.99** U.S.

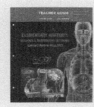

Elementary Anatomy Teacher Guide
Dr. Lainna Callentine

Instructional guide for 36-week elementary anatomy course based on the nervous and the respiratory systems, including weekly calendar, worksheets, activities, and tests focusing on the major concepts.
P 978-0-89051-842-7 **$17.99** U.S.

The Fight for Freedom
Rick and Marilyn Boyer

This third-grade history course introduces readers to 18 heroes of early American history, and some villains as well. Students will learn of God's providential acts with this daily, 34-week curriculum.
STUDENT: P 978-0-89051-909-7 **$29.99** U.S.
TEACHER: P 978-0-89051-912-7 **$14.99** U.S.

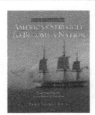

America's Struggle to Become a Nation
Rick and Marilyn Boyer

This fourth-grade course provides a thorough understanding of the foundations of American government. Students learn of the War of Independence through the Constitution in this 34-week study.
STUDENT: P 978-0-89051-910-3 **$29.99** U.S.
TEACHER: P 978-0-89051-911-0 **$14.99** U.S.

Studies in World History 1
James P. Stobaugh

Middle school history that covers the Fertile Crescent, Egypt, India, China, Japan, Greece, Christian history, and more. Begins with creation and moves forward with a solid biblically based worldview.
STUDENT: P 978-0-89051-784-0 **$29.99** U.S.
TEACHER: P 978-0-89051-791-8 **$24.99** U.S.

Skills for Language Arts
James P. Stobaugh

From the basics of grammar to a voyage through classic literature, this 34-week, junior high course lays a foundation for students who are serious about communicating their message. Five instructive lessons weekly.
STUDENT: P 978-0-89051-859-5 **$34.99** U.S.
TEACHER: P 978-0-89051-860-1 **$24.99** U.S.

Studies in World History 2
James P. Stobaugh

Middle school history that covers the clash of cultures, Europe and the Renaissance, Reformation, revolutions, and more. A comprehensive examination of history, including geography, economics, and government systems.
STUDENT: P 978-0-89051-785-7 **$29.99** U.S.
TEACHER: P 978-0-89051-792-5 **$24.99** U.S.

Skills for Literary Analysis
James P. Stobaugh

Equips middle school students to analyze classic literary genres, discern authors' worldviews, and apply biblical standards.
STUDENT: P 978-0-89051-712-3 **$34.99** U.S.
TEACHER: P 978-0-89051-713-0 **$15.99** U.S.

Studies in World History 3
James P. Stobaugh

An entire year of high school American history curriculum in an easy-to-teach and comprehensive volume by respected Christian educator Dr. James Stobaugh.
STUDENT: P 978-0-89051-786-4 **$29.99** U.S.
TEACHER: P 978-0-89051-793-2 **$24.99** U.S.

Skills for Rhetoric
James P. Stobaugh

Help middle school students develop the skills necessary to communicate more powerfully through writing and to articulate their thought clearly.
STUDENT: P 978-0-89051-710-9 **$34.99** U.S.
TEACHER: P 978-0-89051-711-6 **$15.99** U.S.

Principles of Mathematics Book 1
Katherine A. Loop

Focus is on multiplication, division, fractions, decimals, ratios, percentages, shapes, basic geometry, and more, teaching clearly how math is a real-life tool pointing us to God and helping us explore His creation.
STUDENT: P 978-0-89051-875-5 **$34.99** U.S.
STUDENT WORKBOOK: P 978-0-89051-876-2 **$24.99** U.S.

Principles of Mathematics Book 2
Katherine A. Loop

Focus is on the essential principles of algebra, coordinate graphing, probability, statistics, functions, and other important areas of mathematics. Here at last is a curriculum with a biblical worldview.
STUDENT: P 978-0-89051-906-6 **$34.99** U.S.
STUDENT WORKBOOK: P 978-0-89051-907-3 **$29.99** U.S.

Introduction to Anatomy & Physiology: Musculoskeletal System
Dr. Tommy Mitchell

An exploration of the awe-inspiring creation that is the human body. Explore the structure, function, and regulation of the body in detail, our bodies God created that are delicate, powerful, and complex.
C 978-0-89051-865-6 **$17.99** U.S.

High School

American History
James P. Stobaugh

An entire year of high school American history curriculum in an easy-to-teach and comprehensive volume by respected Christian educator Dr. James Stobaugh.
STUDENT: P 978-0-89051-644-7 **$29.99** U.S.
TEACHER: P 978-0-89051-643-0 **$19.99** U.S.

American Literature
James P. Stobaugh

A well-crafted high school presentation of whole-work selections from the major genres of American literature (prose, poetry, and drama), with background material on the writers and their worldviews.
STUDENT: P 978-0-89051-671-3 **$39.99** U.S.
TEACHER: P 978-0-89051-672-0 **$19.99** U.S.

British History
James P. Stobaugh

Examine historical theories, concepts, and global influences of this tiny country during an entire year of high school British history curriculum in an easy-to-teach and comprehensive volume.
STUDENT: P 978-0-89051-646-1 **$24.99** U.S.
TEACHER: P 978-0-89051-645-4 **$19.99** U.S.

British Literature
James P. Stobaugh

A well-crafted presentation of whole-work selections from the major genres of British literature (prose, poetry, and drama), with background material on the writers and their worldviews. High School.
STUDENT: P 978-0-89051-673-7 **$34.99** U.S.
TEACHER: P 978-0-89051-674-4 **$19.99** U.S.

World History
James P. Stobaugh

This study will help high school students develop a Christian worldview while forming his or her own understanding of world history trends, philosophies, and events.
STUDENT: P 978-0-89051-648-5 **$24.99** U.S.
TEACHER: P 978-0-89051-647-8 **$19.99** U.S.

World Literature
James P. Stobaugh

A well-crafted presentation of whole-work selections from the major genres of world literature (prose, poetry and drama), with background material on the writers and their worldviews. High school
STUDENT: P 978-0-89051-675-1 **$34.99** U.S.
TEACHER: P 978-0-89051-676-8 **$19.99** U.S.

ACT & College Preparation Course for the Christian Student
James P. Stobaugh

Your ACT score is key in determining college scholarships and admissions. Prepare to excel with these 50 devotion-based lessons.
P 978-0-89051-639-3 **$29.99** U.S.

SAT
James P. Stobaugh

Written by a certified SAT grader, get insight designed just for Christian students to be well prepared for the test and score higher. Also focuses on spiritual disciplines of Bible reading and prayer.
P 978-0-89051-624-9 **$29.99** U.S.

Building Blocks in Earth Science
Gary Parker

To understand earth science, you must combine the methods and evidences of both science and history. If you also use the "history book of the world," the Bible, you can make sense of the earth's surface.
P 978-0-89051-800-7 **$15.99** U.S.

Building Blocks in Science
Gary Parker

Explore some of the most interesting areas of science: fossils, the errors of evolution, the evidence of creation, all about early man and human origins, dinosaurs, and even "races." High school.
P 978-0-89051-511-2 **$15.99** U.S.

Building Blocks in Life Science
Gary Parker

Teaching high school level science from a biblical perspective. Teachers and students will find clear biological answers proving science and Scripture fit together to honor the Creator.
P 978-0-89051-589-1 **$15.99** U.S.

The History of Religious Liberty Student Edition
Michael Farris

This special edition was designed to be used with the high school course *The History of Religious Liberty*. This student-friendly text has been enhanced with images of relevant people, places, and events!
STUDENT: P 978-0-89051-882-3 **$34.99** U.S.
TEACHER: P 978-0-89051-869-4 **$15.99** U.S.